Monographs on soil and resource surveys

General editor
P. H. T. BECKETT

Ground and air survey for field scientists

JOHN WRIGHT
*Formerly Deputy Director of
Overseas Surveys, Ministry
of Overseas Development*

CLARENDON PRESS · OXFORD
1982

Oxford University Press, Walton Street, Oxford OX2 6DP
London Glasgow New York Toronto
Delhi Bombay Calcutta Madras Karachi
Kuala Lumpur Singapore Hong Kong Tokyo
Nairobi Dar es Salaam Cape Town
Melbourne Wellington
and associate companies in
Beirut Berlin Ibadan Mexico City

© John Wright, 1982

Published in the United States by
Oxford University Press, New York

All rights reserved. No part of this publication may be reproduced,
stored in a retrieval system, or transmitted, in any form or by any means,
electronic, mechanical, photocopying, recording, or otherwise, without
the prior permission of Oxford University Press

British Library Cataloguing in Publication Data

Wright, John
   Ground and air survey for field scientists.—
   (Monographs on soil and resource surveys)
   1. Surveying
   I. Title   II. Series
   526.9   TA545
   ISBN 0-19-857560-2
   ISBN 0-19-857601-3 Pbk

Typeset by Anne Joshua Associates, Oxford
Printed in Great Britain
at the University Press, Oxford
by Eric Buckley
Printer to the University

FOREWORD

This series of handbooks was initially concerned with soil mapping and classification but its scope has been widened to cover all kinds of resource survey. There are now nine volumes in the series and five more in preparation.

Field scientists are well qualified to classify the soil, vegetation, or land use at the point where they stand, or to locate the boundary between two such classes. Unfortunately some of them are less reliable at transferring those observations from the point on the ground to a correctly located point on a map. In lowland Britain things are easy. The 1:2500 or 1:10 000 base maps record so many landmarks that it requires almost deliberate incompetence to record features in the wrong place. This is not the case in parts of the world where maps record fewer landmarks, or featureless landscapes have not got any. In any of these the field scientist must possess at least a little knowledge of survey procedures to locate himself or to lay out an accurate grid of sample sites.

The volume was planned to meet the need. It attempts to equip the field scientist with the basic knowledge which he needs to perform operations himself or to discuss his problems intelligently with a professional surveyor or, as the author points out, to make sure that the latter is not committing some nonsense.

As a mathematician and geographer turned surveyor John Wright brings much experience to the problems of the field scientist:

1934-1939

Expeditions to Iceland, and (over-wintering) Spitsbergen, Greenland, and the Canadian Arctic.

1939-1955

Inspector of Surveys and Provincial Survey Officer in the Sudan, responsible among other things for surveys of the northern desert, cadastral surveys in the Nile Valley, starting an air survey section, and surveys and hydrological investigations for irrigation, and in connection with the proposed Jonglei Canal. During the Second World War he surveyed parts of the Libyan Desert in conjunction with the Long Range Desert Group.

1956-1960

Chief Surveyor in Hunting Surveys and Consultants, responsible for several project surveys in the UK and overseas, for advice on surveys to other specialists.

1961-1977

Assistant Director and Deputy Director in the UK Directorate of Overseas Surveys, responsible for field survey projects and for survey training. He has also been President of the Chartered Land Surveyors, a Vice-President of the Royal Geographical Society and was a founder member of the Photogrammetric and Cartographic Societies.

These varied backgrounds have left their mark on the chapters that follow. I need only add that, having surmounted the initial fears of the unknown, my undergraduate soil surveyors have found John Wright's instructions easy to follow: they have produced adequate maps by chain survey, plane table, and compass traverse. As they come to occupy more senior and responsible positions they will find the later chapters equally helpful when dealing with professional survey organizations.

PHTB

ACKNOWLEDGEMENTS

This book was written in response to an invitation from Dr P. H. T. Beckett, the general editor of this series of handbooks. Not only did he overcome my initial feeling that enough textbooks on survey, both for the professional and the scientist, already existed; but with his teaching experience he helped me considerably in the early stages in arranging the book in its present form. In particular this puts theory and practice hand in hand so that the reader is led gently on from the simpler to the more complicated concepts with plenty of practical examples.

To a large extent, apart from this and from Dr Beckett's continued and meticulous checking of the text for inconsistencies and obscurities, the book has been an individual effort; but I am grateful to Don Proctor, lately President of the British Photogrammetric Society, for reading and checking the three chapters on Aerial Photography and Survey. I am also grateful to Charles Lane, Chief Cartographer at the Directorate of Overseas Surveys, for a similar check of Chapters 11 and 12. Both of them were asked only to check for inaccuracies and mistakes and are in no way responsible for the style or approach to these subjects. I am grateful also for permission to reproduce Figs. 12.3 and 12.4 from a DOS pamphlet on fair drawing.

The Royal Geographical Society agreed to the incorporation in Chapter 7 of material from my article on air photographs published in 1973 in the *Geographical Journal* and later as a pamphlet for use by small expeditions. Mr J. R. Smith gave me useful advice on the adjustment of quadrilaterals using modern electronic calculators and drew my attention to a number of recent textbooks on this and kindred subjects for inclusion in the bibliography.

I am grateful to Meridian Airmaps for permission to publish their remarkable air photograph of Parliament Square which illustrates so dramatically the difference between vertical photographs and maps; and to Wild Heerbrugg for the use of photographs illustrating some of their instruments. I should like to thank Miss Doris Mason who retyped about a quarter of the book from my drafts with her usual accurate interpretation of my handwritten corrections.

Finally I should like to say how this book could never have been written without the help and encouragement of my wife during 42 years of being married to a surveyor. In particular I am grateful for her acceptance of the extra hazards and responsibility which many separations have inevitably involved, and for never trying to persuade me to adopt a more settled and less demanding profession.

West Wittering J. W.
January 1982

Contents

List of plates	xi
1. INTRODUCTION	1
The aims of the book	1
Outline of the book	4
Some common survey concepts and terms	9

Part I. Simple field surveying — 13

2. SIMPLE PLANIMETRIC SURVEYING	15
Introduction	15
A small flat open site (by chain survey)	16
A hilly site (by compass triangulation)	22
A forested area (traversing)	30
3. HEIGHT MEASUREMENT	41
Introduction	41
Spirit levelling	41
Other techniques	55
Plane tabling	58
Photography from ground stations	70
4. OTHER CONSIDERATIONS	77
Errors and specifications	77
The limitations of simple surveys	84
Simple hydrographic surveys	89
Other surveys	95

Part II. Advanced field surveying — 101

5. ANGLE AND DISTANCE MEASUREMENT	103
Introduction	103
Marking the framework	105
Angle measurement with the theodolite	107
Accurate distance measurement	118
6. MORE ADVANCED SURVEYS – PROCEDURES	126
Triangulation with a theodolite	126
Traversing with a theodolite and tape	134
Height measurement	138
Detail survey with a theodolite	144
An outline of computing procedures	152

Part III. Surveying from air photographs — 159
7. THE CHARACTERISTICS AND USE OF AIR PHOTOGRAPHS FOR PLANIMETRIC MAPPING — 161
Characteristics — 161
Relating air photographs to the ground — 164
The geometry of vertical air photographs
The minor control plot — 174
Radial line equipment and ground control — 181
8. USING STEREOSCOPIC PAIRS OF PHOTOGRAPHS — 188
Introduction — 188
The stereoscopic principle — 193
Interpretation — 202
Measuring the heights of objects — 204
9. THREE-DIMENSIONAL MAPPING — 211
The plotting instruments — 211
Rectified photographs, orthophotographs, and mosaics — 221
Ground control for and from air photographs — 226
Locating airborne geophysical surveys — 234

Part IV. The use and production of maps — 237
10. USING EXISTING MAPS — 239
General characteristics — 239
Grids and graticules — 242
Large-scale plans in the United Kingdom — 246
Large-scale plans overseas — 250
Medium- and small-scale surveyed maps — 253
Small-scale derived maps — 254
Maps from satellite imagery — 257
11. PREPARING THE BASIC PLOT FOR REPRODUCTION — 263
How maps are produced — 263
Mapping natural features — 267
Mapping manmade features — 274
Place names and administrative boundaries — 278
12. FAIR DRAWING, REPRODUCTION, AND PRINTING — 382
Fair drawing and scribing — 382
Lettering and printed symbols — 290
Photographic reproduction — 297
Lithographic printing — 302

Glossary — 306
Recommended Reading — 325

PLATES

(Plates fall between pp. 164 and 165 of the text)

1. Using a parallax bar with a mirror stereoscope.
2. The principles of the mechanical type of stereo-plotter.
3. Wild rectifying enlarger.
4. Wild A.8 stereo-plotter.
5. Vertical air photograph of Parliament Square, London.

1. INTRODUCTION

The aims of the book

This book has three aims.

(i) To explain and describe in detail the simplest field survey and air survey mapping techniques so that anyone interested, with access to the simple instruments and materials required, could make a map of a small area.

(ii) To explain, but in less detail, the more complicated and accurate techniques of survey as an introduction to more advanced surveying. In order to practise these himself the reader would almost certainly have to consult more advanced textbooks, and would probably also have to undergo some practical training with the instruments described. He should, however, be able to acquire enough knowledge from this book alone to enable him to understand the techniques used by professional surveyors and cartographers, and as a result should be better equipped to work alongside them in the sort of joint project required for many field investigations − of soils, vegetation, geology or ecology, or indeed any form of geographical or natural resources investigation involving the potential and the actual use of the land.

(iii) To emphasize the strengths and weaknesses of various types of maps and of the techniques used in making them so that more senior scientific administrators will be better equipped to negotiate with professional surveyors and cartographers on a major project and will have some idea of the advantages and disadvantages of the various proposals made to them by survey organizations.

In tackling this task I have tried to fill what seemed to me to be three notable gaps that exist despite the considerable number of textbooks on surveying and mapping that are already available. In this I have been helped by having spent my whole working life in producing maps rather than in teaching others to do so, and also by having worked on both sides of the counter, i.e. for both government and commercial organizations. The first gap is that there are relatively few books that give equal weight to both ground and air surveys or that really explain their similarities and differences; even fewer of them treat the two as equal halves of the same basic technique. This book demonstrates how in some circumstances what are normally thought of as ground survey tasks can be better done from air photographs and vice versa; it also describes in more detail than most textbooks the requirements for the precise identification on air photographs of the ground points used to control their scale and position, which are the essential links between air and ground surveys.

The second gap lies between the simple and advanced text- or handbooks: the former tend to describe simple survey procedures in great detail without going into the theory behind them, which makes them unsuitable for graduates, while the latter go into the theory in great detail with full mathematical explanations. This makes them suitable only for students with a strong mathematical background. Since this book is aimed at field scientists who may have reached a fairly elementary standard in mathematics but nevertheless understand basic scientific principles I try to explain, right from the start, the basic principles by which surveyors work and to show how these affect the design of advanced as well as simple instruments and techniques. In doing this I have tried to use the absolute minimum of mathematical notation or formulae, but rather to express the mainly geometrical principles in physical terms which any trained scientist should be able to follow, even if some of the analogies used may seem rather unusual. Inevitably this has involved some simplification of the theories and principles concerned: for example in the analysis of errors, the design of three-dimensional photogrammetric instruments, and the chemistry of photographic and printing methods. However, I have assumed that once a scientist understands what he is trying to do and why surveyors have evolved over the centuries a particular way of doing it under a particular set of conditions, he will prefer to work out for himself the detailed procedure required rather than be told exactly how to carry out a particular task. This book is definitely not intended to be a handbook for craftsmen who are happy to follow instructions exactly without understanding the reasons behind them.

It should be clear from the opening paragraph that this book is intended to be useful to a field scientist throughout his career and not only in the early stage of working in the field, and this is the third gap that it hopes to fill. Plenty of books exist to guide him in the making of maps to illustrate his particular scientific field, but few describe the more advanced types of survey from the point of view of a scientist turned manager who is working alongside professional surveyors or employing them as contractors. The early chapters are intended for the student or field worker studying soil, vegetation, geomorphology, etc., in which features have to be located on the ground in their correct positions and shapes so that they can be shown on maps. If he wants to compile the map himself, then these chapters on ground and air survey should tell him how to do this, although even at this early stage he will probably need some practical guidance. Later chapters describe the more advanced techniques which relatively few scientists who are not professional surveyors are likely to have the time to carry out themselves. However, it is at this next stage in his career that the field scientist will probably find himself working alongside a survey section responsible for the mapping part of a project, and at some point he may even be the manager in charge of the whole field party. The work of the field surveyors will be laid down by the organization responsible for them and will be directed by the senior surveyor in change, but in almost any large field project problems arise that were not foreseen when it was planned and at the time when the specifications for both the scientific observations and the survey work required to locate and map them were drafted. A field party of this kind may be a long way from the headquarters of both organizations and the more the leader of the scientific party and the senior surveyor understand each other, the better will they be able to work out together the modifications required to the original specifications and methods of working. If the senior surveyor is a professional or graduate he should be able to understand what the scientist is trying to do and know something about all the field sciences for which surveying is required, but if he is not and his education stopped when he left school and his training has been concentrated on surveying alone, then a better understanding by the scientist of surveying principles and procedures will obviously help them to work out together how to solve their joint problems.

The great majority of surveyors, as of other scientist and technologists, are conscientious and good at their job, but it sometimes happens, especially when a survey organization has taken on more, or more advanced, tasks than its own staff can tackle, that extra surveyors are employed at short notice without an adequate check on their competence or reliability. These may even include senior surveyors in charge of field parties. Although obviously he would have to proceed with great caution, a scientist who has some understanding of how survey tasks should be tackled will be alerted earlier to the fact that the survey part of the operation is going seriously wrong and be able to warn his own organization to take this up with those responsible before too much time and money are wasted. At a more senior level still, the scientist is likely to be concerned almost entirely with the administrative direction of projects, mainly from headquarters. Here he may be involved in negotiations with governments or commercial firms who will carry out the survey and mapping sections of the project. Although he should obviously rely if possible on an independent survey consultant to advise him on the merits of various proposals, the ultimate responsibility for deciding which offer to accept will be his, and a proper understanding of survey principles and practice will make it easier to come to the right decision. In particular it may help him to see how the different proposals are inevitably affected by the staff and equipment which the organizations tendering for the work have available. This applies to government as well as commercial organizations; for example, in the simplest and most obvious case an organization not equipped for air survey is likely to recommend doing the whole work by ground survey, while one equipped with all the latest air survey instruments will want to use these as much as possible and recover their cost even when they cannot really produce the results required. Few professional survey organizations will pretend to meet a specification with techniques which cannot do so, but it has been known to happen; it can also happen that the scientist may be pressed to accept some relaxation of the specification itself. It must be remembered that the only satisfactory way to check a map is to repeat a significant part of it, and that designs — for instance of irrigation canals or drains — based on a faulty or inadequate contour survey cannot be implemented without a resurvey: and this can cause an intolerable extra delay and expense in carrying out a project. In major projects involving a survey a consultant should always be employed; this book is not intended to replace the consultant, but it may enable the scientists to understand him.

Outline of the book

The rest of this book is divided into four parts dealing respectively with simple ground surveys (Chapters 2-4), more advanced surveys (Chapters 5 and 6), air surveys (Chapters 7-9), and maps (Chapters 10-12).

INTRODUCTION 5

The second chapter leads straight on to how to make a map of a small area using the simplest forms of field survey, and measuring distances and angles with a tape or chain and a prismatic compass. In the first section distances only are used for a small, open and relatively flat area, then angles only for a larger and more rugged, but still open, area, and finally in the most difficult case of all a combination of both angles and distances is used for an area in which visibility is limited to paths or lines cut through forest or other tall vegetation. In the description of these techniques, which can be carried out using instruments costing only a few pounds, the fundamental principles of surveying are emphasized: first one must work from the whole to the part; secondly, all observations must be subjected to an independent check; thirdly, all survey and mapping operations are in two parts: the establishment of a controlling survey skeleton or framework of precisely fixed and marked points, and then the mapping of detail by less accurate methods from these points. This chapter also shows how the conditions influence the choice of the techniques to be used in each case, and thus emphasizes that when planning a survey the scientists should consider a whole range of techniques and choose the best combination to suit the conditions of the case.

The simple techniques for measuring differences of height and thus for obtaining contours and adding the third dimension to the map are described in Chapter 3. Spirit levelling is described in considerable detail because it is a relatively simple technique but one in which mistakes are easily made; this section also indicates the guiding principles to be followed if assistants are to be employed (directly or through a contractor) in an extensive levelling project in which it would be inefficient and tedious for the scientist to do all the work himself. From this the chapter leads on to the most rewarding, and in teaching terms the most helpful, of all survey techniques — the use of the plane table. This is the only technique in which the map is actually made as the observations are taken, but this can only be done when conditions are suitable and so the last section of the chapter describes how the plane table can be used in the field for the framework survey, with detail to be added in the office or camp from photographs taken from the framework stations in conditions where field time, or weather, or harsh working conditions, or the inaccessibility of the survey stations, all limit the time and mental effort which can be expended at them.

Chapter 4 is a bridge between simple and advanced surveying. First the subject of errors is considered in some detail (but with many simplifications), and it is then shown how simple survey techniques are limited to the production of maps confined to a single sheet or to the surveying of detail based on a more extensive framework which has to be established by more elaborate and accurate techniques. Brief accounts of special survey problems, particularly those associated with surveying the bottom of a sea or lake, underwater features such as wrecks or archaeological sites, and caves, and the measurement of heights of individual objects such as trees or buildings are also included.

More advanced techniques of angle and distance measurement which are necessary for establishing an accurate framework from which several map sheets can be surveyed in detail, in most cases by several different surveyors or scientists, are described in Chapters 5 and 6. This involves particularly the use of the theodolite, pre-eminently the surveyor's instrument, for measuring horizontal and vertical angles. The establishment of proper ground marks and the instruments and their handling, including the more accurate use of steel tapes for distance measurement, are discussed in Chapter 5. The much more expensive electromagnetic distance-measuring instruments which are replacing tapes for more accurate, rapid, and extensive work are considered briefly. The use of these various instruments for triangulation and traverse and the extra refinements required compared with the way these operations were carried out by tape and compass are described in Chapter 6. Few textbooks describe in detail how a theodolite can be used for detail survey in medium- or small-scale mapping or measuring distance from a short or elevated base, so a complete section is devoted to this. The principles of computing are well described in many books, and so they are only outlined briefly here. Nevertheless it is in his failure to follow these basic principles that the sort of senior surveyor described on p. 3 may most easily be seen to be incompetent, however good his arithmetic may be.

The whole subject of surveying from vertical air photographs is dealt with in Chapters 7, 8, and 9. The photographs themselves and the method of relating them to the ground are described in Chapter 7. This is followed by a description of the more complicated though still basically simple means of using single vertical air photographs to build up a survey framework between which detail, but not contours, can be interpolated exactly as on the ground. The radial line principle — that angles at the centre of a vertical air photograph are true — is explained with considerable approximation, and the technique for making a map from a few air photographs using very simple equipment is described. Special equipment is necessary for more extensive areas involving large numbers of photographs, but even this is not very expensive and, since its operation is extremely simple, many scientific departments using air photographs will find it worthwhile. The more complex subject of using air photographs stereoscopically, starting with a detailed explanation of the stereoscopic principle and how its application to pairs of overlapping photographs differs from its normal use, is discussed in Chapter 8. This chapter takes the subject as far as the scientist himself is likely to go in practical applications with its relevance to interpretation and its use in measuring the heights of individual objects such as trees or buildings with the help of relatively cheap equipment.

In Chapter 9 the book enters an area of more interest to those dealing with surveyors as contractors rather than to those making maps themselves, because it describes the elaborate and expensive instruments required to make a completely three-dimensional map which is in general not something that the scientist is likely to attempt himself because of the fairly lengthy training required. However, it has two applications for the field scientist. First, in any extensive project nowadays the mapping is likely to be done in this way by a professional survey organization, and the scientist should have at least a working knowledge of how it is done and of the weaknesses and strengths of the various techniques. Secondly, if suitable air photographs exist of the area of interest, it will be much more efficient, even in a small project, to have detailed maps made of it or of special features such as the geomorphology or the human settlements from the air photographs than to use detailed ground surveys. However, in order to do this the photogrammetrists handling the air photographs must have proper ground control, and it may be a better use of the scientist's time and survey knowledge to establish this than to do the detailed survey on the ground using a plane table or theodolite. Here again many textbooks describe the survey methods required but few also describe the principles of identifying the control points on the air photographs correctly or of where they should be placed, and this book attempts to fill the gap. It also describes the reverse situation in which the photogrammetrist may be able to provide a framework of suitable ground features from which the scientist can locate himself accurately by simple methods using only a tape and compass in order to trace on the ground such features as soil or vegetation boundaries which cannot be seen with sufficient precision or reliability on the photographs.

The characteristics of existing maps are described in Chapter 10 and this points out in particular how the circumstances in which they were made (by ground or air survey or a combination of the two, by compilation from other maps or travellers' reports, or from satellite imagery) affects their accuracy, reliability, and the features that they show. Cartography, i.e. the presentation rather than the observation of data in map form, is dealt with in Chapters 11 and 12. The way in which the scientist should prepare his basic field or air survey plot is described in some detail in Chapter 11, and the ways in which cartographers and printers turn this into the finished map are described in Chapter 12, beginning with the preparation of fair drawings which is described in enough detail for the scientist to attempt at least the simpler stages of this process himself. The printing processes are only outlined since the scientist is unlikely to attempt them himself, but he may be concerned in negotiating with cartographers and printers for the production of maps from his data; this chapter may help him to understand their point of view and the factors which influence costs and time.

The contents of the book have been outlined in this section. However, one important omission must be mentioned and explained. This is that nothing has been said about the use of astronomy in the form of observations of the sun and stars for fixing a scientist's position or obtaining a true *azimuth* or bearing. There are two reasons for this. First, it is a complicated technique and secondly it is now very rarely necessary. Not only are special instruments required in the form of accurate watches and radio sets for receiving time signals, but to calculate one's position with any accuracy requires more complex mathematics than are needed for the kind of surveys described in this book. Special books of tables, such as nautical almanacs or star almanacs, which give the positions of the sun, the moon, the planets, and the fixed stars throughout the year are also required. The formulae used in this book have been reduced to the minimum and include those of simple plane trigonometry only; astronomy in surveying requires an understanding of spherical trigonometry. A complete section would be required to cover the subject properly, and there are already a large number of textbooks which do this at various standards ranging from simple navigation to the nearest mile or so down to precise astronomical fixation to a second of arc or less in latitude and longitude. The second reason for the omission of astronomical techniques is the existence over a large part of the world of adequate maps and survey frameworks, and elsewhere of satellite imagery, from which absolute positions can be deduced not only with greater accuracy than an amateur could expect to achieve by astronomical observations but also in sympathy with the neighbouring features as shown on the existing maps. Because certain approximate assumptions about the shape of the earth are made in astronomical surveying, its observations give positions which may well be out of sympathy with maps and survey frameworks based on astronomical fixes taken elsewhere — these differences can amount to about a kilometre. It is therefore assumed in this book that a scientist preparing a map will either relate it to an existing survey framework or locate and orient it as accurately as possible by the use of existing maps, air photographs, or satellite imagery; the use of astronomical observations would add considerable complications and might produce an incorrect result.

Some common survey concepts and terms

The various technical terms required which form part of survey jargon will be defined in detail where appropriate and they are listed, with brief definitions and reference to the fuller explanations, in the Glossary. However all technical jargon includes, in addition to special words or even neologisms, some examples of common words to which unusual meanings have been given. The burgeoning technology of computers is a good example of this, with special meanings for words like 'hardware', 'menu', 'real time', and so forth; compared with this the longer established technology of surveying has relatively few examples. In this section we introduce the reader to the commoner ones which might cause confusion, whether in field or air survey or in cartography.

Probably the commonest case is that of the word 'detail'. As already indicated, all survey and mapping consists of the establishment of a framework of precisely located points on the ground (sometimes known as the 'control') between which the features to be mapped are then drawn in by more approximate methods or even just by eye. In general terms it is these features which are referred to as detail, but more particularly this term tends to refer only to the plan positions of the features without regard to their height; and the shape of the ground or relief is put in a different category. Thus it is common survey practice to divide the content of a map into detail and contours or, where no contours are shown, whatever method of depicting the relief has been used. Surveyors also distinguish, particularly when using air photographs, between 'hard' and 'soft' detail. The former is almost always man-made and has a satisfactory definite and hard edge such as that of a building, a fence or wall, or a made road; while a footpath, or even more a hedge or a tree, are typical soft detail of which the shape and size may change with the years and seasons while in any case being difficult to define with precision.

'Fix' is another term which has many common meanings (including recently and sadly that of taking drugs), but to the surveyor it means only the precise determination of a point's position on the ground, whether by the measurement of distances and/or angles on the ground, as a result of astronomical observations (as in 'astrofix'), or by measurements on a series of air photographs. All fixes are the result of a series of measurements, some of which are 'redundant', i.e. there are more of them than are actually required to determine the point's position. Comparison of these redundant measurements with the others gives an idea of the accuracy of the fix because they will not all give exactly the same result; when the accuracy of a fix is known it can be described as 'good' or 'poor'.

'Check' and 'closure' are two more terms related to accuracy which are much used by surveyors who, like all good scientists, are suspicious of their results and are constantly trying to determine how accurate and reliable they are. A check may be something quite simple like repeating the measurement of a distance or angle to make sure that it was not misread the first time, but a good check is independent and involves some sort of geometrical closure. The simplest example is that of measuring the three angles of a triangle. One angle is redundant, since two are enough to define the triangle's shape, but the third provides an independent check and closes the triangle. In traversing, where a consecutive series of 'legs' have both their lengths and the angles between them observed, a misclosure is observed when it closes back on itself or is 'controlled' by end points already fixed by some more accurate method.

This brings us to the three terms used for describing errors as described in detail in Chapter 4: 'accidental', 'systematic', and 'gross'. The accidental errors are in theory relatively small, unavoidable, and unbiased. Systematic errors are biased by such factors as using a tape with a different length from that assumed or a compass with a faulty card; these errors are more dangerous because they accumulate and may have quite a large cumulative effect. Gross errors or blunders are mistakes which have an effect much larger than the accidental ones. Errors are also described as 'absolute' or 'relative'; in the first case a position is known with an absolute error in terms of a grid or latitudes and longitudes, while in the second case its position is known with an error relative to the features round it. For most purposes the relative accuracy is of more practical interest than the absolute accuracy. For example, unless an instrument such as a theodolite or tape is used, a map user will not notice absolute errors at all and will hardly notice relative errors greater than about 1 part in 50 of distances to neighbouring objects, which is about as accurately as he could pace them. It is this difference between the two types of error which is one reason for separating the survey process into the two parts of precise fixing of the points of a framework with small absolute errors, followed by fixing detail from them with small relative errors which can be done much more easily and cheaply with simple equipment and techniques.

In the use of air photographs two common terms are 'identification' and 'interpretation'. To a surveyor identification means only the precise determination on the air photograph of the position of a feature on the ground; this is usually done on the ground by studying the air photograph and comparing the appearance of the feature of interest with the other features nearby. It could be the corner of a house or some other suitable hard detail, or soft detail like a path intersection or a small tree; and a gross error could be made by identifying the wrong tree or house. Interpretation is a less precise matter in which the nature rather than the position of the feature being viewed is determined and it is placed in a category such as a particular class of building, tree, or road, or an area of soil, vegetation, or rock of a particular kind. While a single photograph can be used, it is more likely in both cases that two overlapping ones will be viewed stereoscopically in order that the feature can be seen in relief with all three dimensions visible; obviously this makes both identification and interpretation more precise and reliable. This leads on to an important concept in air survey which is the 'overlap', i.e. the area common to two successive vertical air photographs taken along a flight line. Normally this is about 60 per cent of the area of each photograph, and there is also a smaller lateral overlap with the photographs of adjoining parallel flight lines. By this means every part of the area is visible from at least two air photographs and so can be viewed stereoscopically; thus the overlap is the unit of area which really matters rather than the area covered by a single photograph.

In cartography, which is the presentation of data in map form, three common terms with meanings different from the normal ones are 'plot', 'copy', and 'reproduction'. Plot does not mean a conspiracy but the map as it comes from the field or, more commonly, from the air survey laboratory and before it has been turned into fair drawings suitable for 'reproduction'. This has no biological meaning in cartography, but describes all the processes, both photographic and mechanical, by which the original plot is finally reproduced in several copies on paper as a map. 'Copy' is confusing because to a reproduction operator it means the original from which he makes negatives or copies in other forms. Other terms, most of which will still be words familiar in some other form to the reader, will be defined as they appear but we hope that this brief introduction to some of the commoner special uses of ordinary words will help the reader to avoid misunderstanding and confusion.

Two final points can conveniently be made here. The first is that although the scientist or reader is referred to throughout as 'he' this is not intended to suggest that all scientists or surveyors are male; it is simply a matter of convenience backed by a sound legal precedent since a learned judge once ruled that in legal documents, where the same uncertainty of meaning might arise, 'the male embraces the female'. The second point is that readers will find a fair amount of repetition in this book, particularly of important axioms and principles and those which are not, in the opinion of the author, emphasized sufficiently in existing textbooks. This is a book primarily for the reader who wants to know *why* as well as *how* particular survey procedures are used, and the basic principles behind them cannot be emphasized too often. Moreover it is not likely to be read straight through, but selectively as a textbook or work of reference, so the repetitions will be less noticeable. May I conclude this introductory chapter by hoping that the reader who is stimulated by this book into going further into surveying then he originally intended will find it as fascinating an occupation as I have, and that all readers will find it as interesting to read as I have to write.

PART I
Simple field surveying

2. SIMPLE PLANIMETRIC SURVEYING

Introduction

In this chapter we introduce some simple survey tasks and draw from them the basic principles on which all surveying and mapping are based. The theory, and to some extent the practice, of more advanced techniques are described in later chapters, mainly in order to help the reader discuss his requirements with professional land surveyors and to appreciate their point of view. Three different kinds of site are considered, each of which requires a different technique because of topography and vegetation cover, or the size of the area, or the features to be mapped. The field scientist is assumed to have the following equipment available:

(i) 100 ft or 30 m plastic or linen tape and 'arrows';
(ii) a small hand prismatic compass;
(iii) a flat drawing board of dimensions about 60 cm × 40 cm covered with good quality drawing paper;
(iv) a ruler or scale at least 30 cm long reading in centimetres and millimetres, and a circular protractor;
(v) binoculars, half a dozen striped ranging rods, two dozen wooden pegs and a mallet, hard and soft pencils and a sharpener, an india rubber, a fine pen and waterproof ink, and a note book for recording.

The three sites discussed are as follows, in order of increasing difficulty:

(i) a small site of dimensions about 100 m × 100 m on relatively flat ground with few trees or bushes, for example an archaeological site or a system of agricultural trial plots;
(ii) an open hilly site about 1 km square where, for example, a series of agricultural terraces or ancient fortifications or soil erosion on steep slopes are to be mapped.
(iii) a heavily forested area, also fairly hilly, in which an irregular system of paths and tracks and stands of different trees are to be mapped.

A small flat open site (by chain survey)

The significant features of this site are its flatness, openness, and small size, because this means that it is easy to measure distances between points on the ground by simply stretching the 30 m tape between them. This technique is called *chain surveying* because a chain with special links which lasted longer than the tapes then available used to be employed; now these are made of a plastic such as fibre glass which is much tougher and less liable to fray or rust, or break when kinked than the old linen or steel tapes. The first step, as in any survey, is to gauge the size and shape of the area so that the appropriate *scale* and its location on the drawing board can be determined (Fig. 2.1(a)). This can be done simply by inspection or by identifying it on a large-scale map such as the UK 1/2500 scale where it will be about 4 cm square, i.e. large enough to show its shape. The next step is to choose, either along one boundary or preferably somewhere through the middle, the longest possible straight line of which the ends are intervisible with a surface flat and open enough for the tape to be laid straight along it. In the diagram this base line has been taken as running through the middle of the site from east to west. Points E and W are chosen at each end from which, as far as possible, similar relatively long flat unencumbered lines radiate out, and a peg is hammered in at each of these two points leaving only one inch projecting; ranging rods are erected at each end near these pegs.

The length WE is then measured with the tape, starting first from the western end with the tape zero on the peg; one man holds it there and directs the other to walk towards the ranging rod at the eastern end and then stretch the full length of the tape along the line and put in an *arrow* (a long skewer with coloured tape attached) at the 30 m mark. Simultaneous accuracy at the two ends is confirmed by shouting. The tape is then moved forward until the back man can put the zero mark against this arrow and a second one is put in again at the 30 m mark. This process is repeated a third time and the short distance to the eastern peg, say 5.13 m in this example, is then measured to the nearest centimetre making a total of 95.13 m. The whole distance is then measured in the opposite direction in the same way, but starting with the full 30 m from the western point so that the arrows are not put in the same positions as on the first measurement. Provided that the difference between the two measurements is only a few centimetres, the mean of the two can be accepted as the true distance. In order to

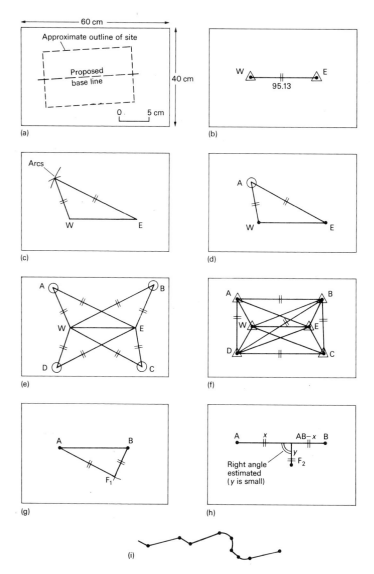

Fig. 2.1 Chain survey: (a) sketch map of the site; (b) base line measured and plotted; (c) distances from E and W to A plotted as intersecting arcs; (d) position of A fixed by lengths WA and EA; (e) positions of A, B, C, and D fixed provisionally; (f) check lines measured and fix A, B, C, and D fixed finally; (g) fixing a detail point from two fixed points by distances; (h) fixing a detail point by distance and offset (AB − x is measured as a check); (i) choice of detail points along a feature.

18 SIMPLE PLANIMETRIC SURVEYING

fit the map of this site onto a sheet of paper of dimensions 60 cm × 40 cm this length of 95.13 m must be reduced to about 40 cm, and a scale of 1/250, where 100 m on the ground is 40 cm on the map, will do this nicely. Each millimetre on the map then represents 25 cm on the ground, so we can plot to the nearest 10 cm on the ground. The base line will be 95.13/250 = 38.05 cm on the map, and this is drawn straight across the middle of the paper with the ends marked by points inside small triangles (Fig. 2.1(b)).

The first survey principle is to work from the whole to the part and not vice versa. Therefore the next step is to obtain a better idea of the outline of the area to be surveyed so as to check at an early stage that we have chosen the scale and the positioning of the base line correctly in order that the whole area will fit onto one map sheet. We therefore next choose the four corner points and mark them with pegs, selecting points near the real corners of the site, and preferably outside them, such that they are intervisible from each other and that the tape can be laid between them and also between them and the two ends of the base. The lengths to each of the corner points A, B, C, and D are now measured from E and W the two ends of the base. In each case, as shown in Fig. 2.1(c), we plot them by setting out on the board arcs at right angles to the measured distances from E and W and marking where they intersect. A large pair of dividers can be used for this, but it is perfectly possible to do it as shown in the diagram using the scale, if this is long enough, or a piece of stiff paper on which the distance has been marked along one straight edge. The intersection of the two arcs should be pencilled lightly and both distances checked; it is then circled as a provisional point and the arcs are rubbed out (Fig. 2.1(d)). All four points are drawn in this way (Fig. 2.1(e)). Clearly the same technique could be used if the original base was not near the middle of the site but along one of the sides.

Why do we show the points as provisional? The answer is because we could have made a mistake in reading the tape or in counting or plotting the number of tape lengths. Thus we must now apply the second survey principle: *always make an independent check whenever possible.* The easiest way of doing this is to measure the four sides AB, BC, CD, DA (Fig. 2.1(f)); if any of these present obstacles we can measure the diagonals AC or BD instead. Provided that the measurements agree with the distances on the map to about 25 cm on the ground we can take the plotted positions as confirmed, and we show this by changing the circles round them to small triangles and inking them in. We now have an accurate *framework* of points, the positions of which are correctly shown on the map and are marked precisely (to about 1 mm on the map) and unambiguously on the ground by pegs which have been driven in sufficiently deep to be difficult for even small boys (the curse of surveyors all over the world) to pull out.

By this time the reader may be beginning to wonder when we are actually going to start mapping some of the features on the ground, since so far all we have achieved is six pegs in the ground and six points on the map, none of which relate to the features in which he is interested. The answer is that all survey work must be based on a framework in the same way that a modern office block depends on its steel framework. As in this example, the framework is also invisible most of the time but it is nevertheless essential to preserve the shape of the visible part of the structure. Having produced a framework which covers the area, we can now fix any point in the area by measuring the two distances to it from any two of our framework points as shown in Fig. 2.1(g). Alternatively, as shown in Fig. 2.1(h), we can fix a point near one of the straight lines of the framework by measuring the distance y from it to the line, and the two distances x and $AB - x$ along the line to the foot of this perpendicular. We measure both to provide a check. Clearly we cannot fix every point along every hedge, ditch, or soil boundary in this way; it would take too long. What we do is to select points (which we call *detail* as opposed to *framework* points) along the feature such that the rest of it can be interpolated or sketched by eye between them without causing visible errors on the map, say not more than 1 mm. Thus if the feature has a long straight section we only need to fix the points at each end of this; but where it is curved they will have to be closer together; all this is shown in Fig. 2.1(i).

It will be clear that once the framework points have been established by careful and accurate measurement and subjected to independent checks to guard against unacceptable errors, the detail survey can proceed anywhere by quicker and less accurate and reliable techniques. Without the framework a detail survey done in this way would soon accumulate errors which would not only mean that the features were wrongly positioned on the map but that when two different sections of detail survey started from the two ends of the map (possibly by two different scientists) met in the middle they would not agree. We have now established four fundamental principles of surveying:

(i) Survey from the whole to the part, and from outside in.

(ii) Establish first a framework of points which must be precise (i.e. small pegs and not stones or indefinite natural features), unambiguous (one stone is very like another), and permanent (stones can be moved) or at least permanent enough to last for the duration of the survey and its projected future use.

(iii) Either repeat each framework measurement at least once (as in the base line) in such a way that the same error is unlikely to be repeated twice, or better, provide a completely independent check as with the four other points (and of course the two base points also) by measuring check distances which must agree with the distances scaled off the map if the earlier work is correct.

(iv) Having established an accurate and reliable main framework which contains the whole area, a secondary framework of detail points can be fixed by rougher and quicker measurements along each feature to be mapped, and finally we can rely on the estimation of distances and shapes to sketch in the features between them. Detail points are fixed by a single form of measurement and are not independently checked; a check against mistakes is provided by their relationships to each other when interpolating between then.

All field survey relies on a similar sequence, but if we cannot use direct measurement (and over longer distances without sophisticated and expensive equipment it becomes very difficult) we can sometimes manage by measuring only one or two distances and deriving all the rest by using bearings or angles.

However, before we leave this site we should do something that is normally required by archaeologists, i.e. establish a *grid* over it so that every artefact recovered can be located in a replica of this grid and labelled with its position for future reference. This brings out the fact that surveying is in two parts: putting on paper what is on the ground, which is mapping, and putting on the ground what is on paper, which is *setting out*, whether it is the grid required by the archaeologist (or by the agriculturalist in plots required for trial planting) or in more extensive projects a canal system designed on a contoured map, a carefully designed dam or bridge, or a housing estate or new town. In our case the grid has first to be designed and drawn on the map which can be done by drawing the largest possible square or rectangle using a ruler (or a straight-edged piece of paper) and using arcs to fix the other two corners with the lengths of the sides and diagonals. In a square these diagonals will be $\sqrt{2}$ (1.414) times the side lengths; in a rectangle they must be computed by Pythagoras' Theorem from the square root of the sum of the squares of the two sides, e.g. a 3, 4, 5 triangle can be used to set out a rectangle. The side lengths should be chosen as multiples of the sides of the required grid squares, and once the main rectangle is drawn each side is divided into the grid square sides and the lines are drawn across parallel to the main rectangle sides. The relationship between the grid corners and the framework points is scaled off the map and each main grid corner is then established on the ground using the two arcs method in reverse, i.e. scratching the arcs on the ground, seeing where they intersect, and then checking once more the actual distances from this to the two framework pegs. These corner points are the fundamental framework of the whole grid, so the distance to a third original framework point should also be checked with this scaled distance from the map and all four sides of the grid outline should be measured. Once the four main corners are established the main sides can be set out and then subdivided into individual square sides, and the internal square corners set out from these. Once again we have worked from the whole to the part and checked each stage before starting on the next.

Once the grid corner pegs have been established, the importance of the original framework pegs might be thought to have disappeared, but this is not the case. As the excavation proceeds many of the grid corner pegs will have to be removed and replaced at a lower level. If this is done from each other in turn the grid will in time become inaccurate; for example, if two of the original pegs are replaced from two others and later these themselves are replaced from the replaced pegs, the errors will accumulate and the time may come when we have to go back to the original framework to re-establish a considerable part of the grid in its exact original position. As we shall see in later chapters it is a common weakness of non-surveyors, and even regrettably of some surveyors, to underestimate the length of time for which a survey, and particularly its framework can be used. They waste the time and money spent on the original framework required for later setting out or larger-scale surveys by failing to spend the relatively small extra effort required to establish really permanent marks; if these disappear the whole labour of the framework survey is wasted and it has to be repeated.

A hilly site (by compass triangulation)

For the second task I have chosen an open, i.e. not heavily forested, site with features to be surveyed on the slopes; in this country they could be small fields or soil boundaries or rock outcrops, and overseas they could be terraces, or slopes steep enough to cause concern about soil erosion. At this stage no techniques for measuring heights or slopes are being considered, and we are simply endeavouring to make an accurate *planimetric* map, i.e. without contours or heights, of what is on the ground. The site is assumed to be considerably larger than the previous one with an extent of about 1 km square, and this, and its hilly nature, make it impracticable to measure long distances between the corners of the site with a tape as we did in the previous case. In fact, measurement of the distances will be impossible except along one or two lines specially chosen for their flatness. Two possibilities are considered: one a valley with steep sides (Fig. 2.2(a)) and the other a domed hill top (Fig. 2.2(b)) of the sort on which ancient forts and

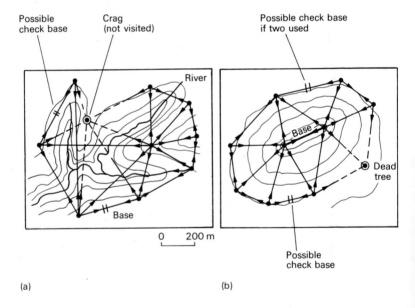

Fig. 2.2 Triangulation: (a) framework for a river valley; (b) framework for a hill top. ⊙→ - - -, rays observed from one end; ⊙→←⊙, rays observed from both ends.

SIMPLE PLANIMETRIC SURVEYING 23

and defence works are found in this country. In the first site we shall obviously obtain a series of good views along straight lines across the valley and down into its bottom from points high up on the sides, while in the second we shall not be able to see across it but we shall be able to see at least the tops of two poles stuck into points on the domed summit from points all round the slopes of the hill. The main characteristics of both sites are therefore great difficulty in measuring distances but plenty of intervisible points with comparatively long straight lines in the air between them. Therefore in both cases we look to the compass rather than the tape to solve our problem.

The simplest instrument for measuring angles, and one with which most geographers and other field scientists are familiar, is the hand *prismatic compass* (Fig. 2.3(a)). It consists of a compass card containing one or two magnets which is engraved in degrees from 0° to 360° round the edge in an anti-clockwise direction. The numbers are reversed because they are designed to be read through the prism which includes

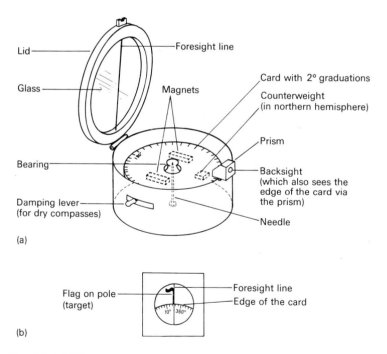

Fig. 2.3 (a) The prismatic compass. (b) View through the back sight; the bearing is 005° since the graduations are each 2°.

a mirror. The card is numbered in the anti-clockwise direction because it remains fixed, with 0° pointing to magnetic south, while the compass has to be turned to point along different bearings in turn. The centre of the card rests on a fine needle point fixed to the bottom of the compass case so that the card can always turn to the nearer of the two magnetic poles. Since these are deep in the earth in northern Canada and the Antarctic, the magnet tries to dip towards the nearer one in high latitudes and has to be counterbalanced by a small weight fixed to the opposite side of the card; thus a compass designed to work in the northern hemisphere may require adjustment for work in the southern. The best prismatic compasses are filled with a liquid which damps their oscillations, but dry compasses contain a damping lever which lifts the card off the needle and brakes it. The lid hinges at the opposite side to the prism and includes a circular window with a line across it which acts as a foresight when it is open and vertical, while the prism holder includes a V backsight. The observer looks through the eyepiece of the prism and sees simultaneously the target, the vertical line of the foresight, and the readings on the card (Fig. 2.3 (b)). In a hand compass the card is graduated in even degrees and can be read by estimation to half a degree. Care must be taken not to wear metallic objects on or near the head or the observer will come back, like my uncle in the war (in his tin hat), with all his bearings reading south.

Most readers will know that the compass provides a *magnetic* bearing not a true one, and not an angle (except from magnetic north or south), although equally obviously if the bearings to two different objects are observed the difference is the angle between the rays to them. Most readers also know that magnetic north does not coincide with true north (and near the magnetic poles may differ by over 90°); it may also differ in different parts of an area especially if the rocks contain ferrous ore bodies. Thus, whenever we can do so it is better to use the compass for observing angles rather than bearings, since the angles will not be affected by changes in the magnetic variation from one point to another. The other advantage of working with angles is that whenever a simple figure like a triangle has all three angles observed an independent check is available since these should add up to 180°. A preliminary check can also be obtained, of course, by comparing the forward and backward bearings between two points, although as explained they may not agree if there are local anomalies in the magnetic variation.

Having practised observing with the compass the scientist is now ready to start actually surveying. As with the first site we assume that he has already walked round the site and obtained an idea of its shape and size and sketched this on his drawing board. Since it is a good deal larger than the previous site the scale will have to be considerably smaller; to accommodate 1 km × 1 km on a 60 cm × 40 cm map will require a reduction of 1/2500 when 1 mm on the board will represent 2.5 m on the ground. Positions for the main framework points have to be chosen as on the previous site, but the criteria are now different since intervisibility is the only one and there is no requirement for flat ground between them. With a valley as shown in Fig. 2.2(a) the layout is fairly straightforward and obviously strong, but with a hill as shown in Fig. 2.2(b) some trial and error may be required before the two points on the hill top are satisfactorily sited so as to give views down what may be convex slopes to the points round the perimeter; and one or two of these perimeter points may have to be further out in order to be intervisible with at least one of the points on the summit. Since we do not have to visit all the points with one end of the tape, it is possible to include a few 'up points', i.e. intersected points which are inaccessible (like a church spire, a mosque minaret, a dead tree or a rocky crag) provided that they can be seen from at least three accessible stations. Examples are shown in Figs. 2(a) and 2(b). Following the usual survey convention the rays to 'up' points are 'pecked' at the end nearest to them since there will be no sights back from them. It will be appreciated that while angles give shape to a figure or map, at least one side of one figure must have its length measured to give the size of the whole framework. If the framework is not too large one of these will be enough, so that on both diagrams a line on relatively flat ground has been chosen and marked as *base*. If the framework is extensive or weak, i.e. if it includes several angles under 30°, then two bases should be included at opposite sides of the area to act as an independent check. These bases are measured exactly as described in the previous section, but since they are longer (perhaps 300 m) they will require the tape to be laid down several times and extra care must be taken to ensure that the number of tape lengths is not miscounted. One way to do this is to use only 25 m of the tape on the second measurement, with the 25 m graduation unequivocally marked with a piece of string or coloured tape.

The chosen stations or points of the framework must be 'permanently' marked (except of course the up points) with pegs as on the

26 SIMPLE PLANIMETRIC SURVEYING

previous site, and they must also be *beaconed* with ranging rods to which flags are attached. There should be enough flags to ensure that all stations visible from any one station can be marked when it is visited. It will save time to have one or two assistants and to work out a programme of observation and of beaconing so that they can move the beacons to the new points required as the observer moves across the area. After drawing the kind of observation diagram shown in Figs. 2.2(a) and 2.2(b), it will pay to decide which rays are to be observed, since observation of them, especially on the first site where many are intervisible right across the area, will add little to accuracy while causing confusion. The framework should be built up of as many and as large *well-conditioned* (i.e. with no angle less than 30°) triangles as possible. Quadrilaterals with both diagonals observed will give added strength to the framework. Strength and accuracy have much in common; most people have an instinctive feeling for the kind of simple structure of straight lines which will be rigid and strong, and exactly the same intuition should be applied to the design of a triangulation framework.

When all the bearings shown in the Figs. 2(a) and 2(b) by arrows along the rays have been observed, the next stage is to extract them from the notebook in two ways: by stations and by figures, and then to analyse them and see what we have got. Figure 2.4 shows a relatively simple triangulation scheme comprising one *braced quadrilateral*, i.e. including its diagonals, and a single triangle. For the sake of simplicity the angles are observed to 1°. If we look carefully at the results we can see that three things have happened.

(i) All the triangles *close* within 1° or 2°, which is reasonable, except for ABE where the misclosure is an unacceptable 7°. Something has gone wrong here.
(ii) The bearings and back bearings, i.e. A to B and B to A, all agree except that all the bearings from C seem to be about 15° greater than they should be compared with those in to C, i.e.

bearing DC is 30°, but CD is 226° instead of 30° + 180° = 210°
bearing AC is 47°, but CA is 242° instead of 227°
bearing BC is 117° but CB is 312° instead of 297°.

Clearly, therefore, there has been an anomaly at C due to some magnetic body nearby giving a variation 15° or 16° greater than

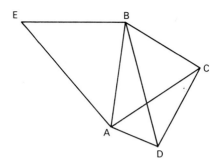

Fig. 2.4 Observations and checks on a triangulation.

Observations				Triangles		
Object	Bearings (deg)	Angles	Values (deg)	Triangle	Vertex	Angles (deg)
Station A						
B	03	BÂC	44	△ABE	A	67
C	47	BÂD	120		B	85
D	123	EÂB	67		E	35
E	296	CÂD	76			187
					Error +07	
Station B						
E	269	CB̂D	44	△ABC	A	44
C	117	DB̂A	23		B	67
D	161	AB̂E	85		C	70
A	184	AB̂C	67			181
					Error +01	
Station C						
D	226	BĈA	70	△ACD	A	76
A	242	AĈD	16		C	16
B	312	BĈD	86		D	88
						180
					Error 00	
Station D						
A	302	AD̂B	39	△BCD	B	44
B	341	AD̂C	88		C	86
C	30	BD̂C	49		D	49
						179
					Error −01	
Station E						
B	80	BÊA	35	△BAD	B	23
A	115				A	120
					D	39
						182
					Error +02	

28 SIMPLE PLANIMETRIC SURVEYING

> everywhere else. If we plotted the bearings directly instead of deriving the angles we should obviously be in difficulty since other plotted bearings would not agree.
>
> (iii) The bearings and back bearings EB and BE do not agree (EB = 80° but BE = 269° when it should be 260° or alternatively EB should be 89°). The other bearing from E, to A, only differs by 1° from its back bearing, so there does not seem to be an anomaly at E. One of these bearings must therefore be wrong, either through a misreading or because the flag was incorrectly seen or sited, i.e. not over the peg. Since all other rays into B were confirmed as correct, our first suspicion falls on BE, and remeasurement makes it 262° and the angle AB̂E 78° which then closes the triangle ABE exactly without affecting any other angle.

The main point made by this example of analysing a set of observations is that they must be set out clearly and in an organized way without trying to see how they come out before they are complete and the necessary copying and arithmetic has been done *and checked*. Experience shows that copying is at least as fruitful a source of errors as the actual observations, and the observer should check his copied results very carefully or, better, read them to a colleague who ticks them off; in this case he should deliberately misread one from time to time to check that his colleague has not gone to sleep — or nearly so. One of the rewards of surveying is in seeing how checks come out, but the temptation to find this out too quickly, e.g. at a station when observing the third angle of a triangle, must be resisted or observations and computation will be mixed up with fatal results. Another rule is never to rub out a faulty observation; it should be cancelled by a line and the correct result written over the top. Having done the arithmetic and checks and confirmed that all the observations are correct (after any necessary re-observation) the results can be plotted. Obviously the base must be plotted first in approximately its correct position on the paper; the rays from it can then be plotted until the whole framework has been drawn, leaving the intersected points to the last after completion of all the fully observed figures. All angles should be checked as they are plotted, for mistakes can also occur in plotting. Normally the map will be oriented to magnetic north by setting out the base by the bearings along it, and true north can then be shown by an arrow at the side of the map; if the magnetic declination is large then this can be taken from the best available local map and the map oriented to true north and south.

Having established a precise and reliable framework, the next step, as in the previous survey, is to establish detail points along the features to be mapped. The distances will now be shorter than between the framework points but may still be a 100 m or more and over rough ground, so we continue to use bearings or angles. Two bearings from the framework points to each detail point are normally sufficient but three should be used whenever easily possible. For instance, features down in the valley bottom can obviously be mapped in this way from points above them. A different technique, called *resection*, must be used instead of intersection for features that cannot be recognized from a distance such as the indefinite crest of a round hill. For *intersection* the observer stands at known points and takes rays *into* the point to be fixed; in resection he stands at the point to be fixed (whose position is *not* known) and takes bearings *out to* at least three fixed points of the framework. Drawing the back bearings will give him his position, provided that there are no local anomalies. His three points should if possible form a triangle round him; if they are on one side only they should be widely separated and the middle one nearer than the other two. If there is a magnetic anomaly where he stands the rays will not meet at a point, and the best remedy is to draw the three rays on a piece of tracing paper, move it until the rays fit over the three points, and then prick through the position of the resected point. Up points are more valuable in resection because, being tall, they are often visible over much of the area when some of the low-lying framework points are not. Where three framework points cannot actually be seen from a detail point on a feature, another point where they are visible should be chosen near it and the actual detail point on the feature can then be fixed by a simple bearing and measured distance from this. If it is found that extra framework points are required, these can easily be added by either observing the three angles of a triangle including a new point and two existing ones or by intersecting it with at least three rays. New framework points should not be fixed by resection. As far as possible detail points should all be fixed from framework points and not from each other, but when fixing a long feature by a series of resections a ray can be taken forward from each resected point to the next one and then fixed by two outward bearings to framework points which will give a check. At times the distance from a nearby framework point or from another detail point can also be used to fix oneself or as a check. With practice the scientist will begin to see how there are various combinations of bearing, angle, and distance measurement which can be employed to solve each problem. When he has done this he is ready to tackle the more difficult site described in the next section where he will have to use traversing.

A forested area (traversing)

In the previous section we saw how a combination of distance and bearing or angle, rather than either alone, could be used to fix a point. In an area where visibility is limited by trees, as in the third site, this is the only way in which we can solve the survey problem because we cannot see across it either to measure bearings between its corners or to lay a straight series of tape lengths. We have to use the tracks or paths which run round or through it, but the frequent bends in these mean that our observations are limited to relatively short sights and measurements of length. However, even in a mountainous forest area these routes should have reasonable slopes (say under 1 in 10) and be fairly smooth, so we should be able to measure distances along them provided we take care. The substitution of a series of shorter lengths with angles between them for a long straight line between two non-intervisible points of a framework is known as a *traverse*, and it has always been thought of by surveyors as a technique to be avoided as much as possible. The reasons for this will become quite clear, particularly when distances have to be measured rather laboriously with a 30 m tape which has to be laid down two or three times along each leg of each traverse.

A diagram of the area to be surveyed is given in Fig. 2.5(a) with the salient points marked by letters. Clearly it is more difficult in this case to obtain even a rough idea of its size and shape by walking round it; however, in a developed country there should be maps which indicate this even if they are out of date, and in less developed ones there may well be at least a small-scale air photograph. In the United Kingdom, for example, we have a complete series of 6 in to 1 mile (1/10 000 scale) maps on which all forests are shown and in rural areas an almost complete series of 1/2500 maps, so it should be possible to obtain a sufficiently good idea of the shape and size of the area to decide on the scale of the map and where the road side of it should fall on our plotting sheet and board. Even the road is not

SIMPLE PLANIMETRIC SURVEYING 31

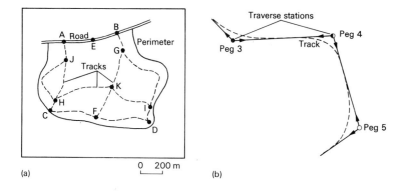

Fig. 2.5 Traversing: (a) traverse layout for a forested area; (b) traverse stations at corners.

straight and the tracks have a number of bends, so that the longest straight line on these is less than 1 km and is often nearer to only 100 m. What we have to do is to break these irregular connecting lines, which will form our framework, into the smallest possible number of straight lines or *traverse legs* along each of which we can take a bearing and measure a straight line distance. The first step is to decide which paths to follow and to choose the traverse stations at the end of each leg. Assuming that from our map or air photograph, or with local assistance, we can pick out and follow the main perimeter paths ACDB, we walk along these and decide where to put our stations. Every traverse station requires the measurement of at least two bearings with all that this involves in causing errors, so we want as few stations as possible and it will pay to put them on the outside of each bend as shown in Fig. 2.5(b). It will also pay, if we can and are allowed to do so, to go several metres off the track and clear the two tangents to the curve if by so doing we can go round the corner with one station instead of using two or more. This will not reduce the amount of taping to be done where of course errors can also arise, but these are normally smaller than those caused by bearing errors using a hand prismatic compass which is all we have for angle measurement. Having selected and if necessary cleared the lines to our stations we hammer in a peg at each and number it in pencil.

It will be noticed that we included one station for the bend in the road — again probably off to one side to give a view along both straight sections — and also pegs at each of the junctions with internal paths; this is to hold down the ends of the traverses which we shall also be running along them. We start observing along the two legs of the road traverse AB and take extra care on this because with only one angle we should obtain a particularly good result and can use it as a base for the whole framework; it is assumed that the slopes along it are quite small. Taping along it and along the other traverse legs is the same as in the previous surveys, except that because the distances are longer and there are more of them we are more likely to make mistakes through boredom and the consequences of doing so are much more serious. In the previous surveys any mistakes were easily discovered: a taped figure did not close or a bearing and back bearing did not agree. Here, if we make a mistake in taping or in measuring the bearing of one traverse leg, the result will not appear until two or three traverses complete what is called a circuit and have been plotted. We then find that they fail to arrive back at the starting point by an amount which is so large that we know that we must have made a mistake somewhere — but where? That is the difficulty. Thus all distances must be carefully taped in both directions and we must be careful not only to measure each tape length (or 'bay') correctly but to the same point; we must also ensure that we do not miscount the number of tape lengths or misread by 1 m or 10 m and that we do not misread our compass bearings by $10°$ or in circumstances when metal is affecting the compass. The best technique for taping is for the forward man to put in an arrow or long skewer with a piece of red rag attached at the end of the tape when it is stretched; the back man then puts the zero of the tape against this when he reaches it and picks up the arrow. If the forward man starts with a known number of arrows they compare notes at the end and check on the number of tape lengths. Ideally a tape of different length and with different units should be used for the second measurement, or the same tape can be used for both with metres on one side and feet on the other; in this way a truly independent measurement is made with completely different results, and so repetition of the same misreading or miscounting is unlikely. Bearings and back bearings should be compared, as should forward and back tapings, and any discrepancies resolved on the spot by re-measurement. Why traversing is to be avoided if possible is beginning to emerge — it is tedious and mistakes are very easily made because of the volume of the observations and their repetitiveness.

SIMPLE PLANIMETRIC SURVEYING   33

In the two previous sites all the distance measurement was along fairly level ground; the first site was level anyway (or it would not have been suitable for a chain survey) and in the second we could choose the one or two triangle sides which we used as bases to be as flat as possible. Here we have no choice, and since most forest areas are hilly we may expect to encounter slopes of up to at least 1 in 10 (about 6°). Examination of Fig. 2.6(b) shows that if these slopes extend for a

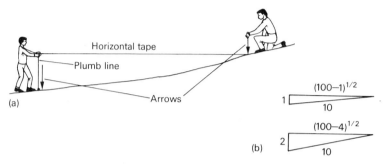

Fig. 2.6 Chaining on slopes: (a) step chaining; (b) slope corrections.

considerable distance the length taped along the slope will be appreciably greater than the true horizontal distance to be represented on our map. By Pythagoras' theorem, a 10 m length of tape on a slope of 1 in 10 will give a horizontal distance of $\sqrt{(100-1)} = \sqrt{99} = 9.95$, so the horizontal distance is 5 cm short of the measured distance. If we have no way of measuring slopes we cannot compute this error and we must use *step chaining* as shown in Fig. 2.6(a) in which the tape is pulled horizontal and the point vertically below the down slope end is marked with a plumb line or by dropping a stone and marking where it falls. This error does not seem very large but it is always positive and systematic; the horizontal distance is always shorter than the measured one (if we tape along the slope) whether we are going up hill or down. If therefore the slope is 1 in 10 all the way along AC, which is about 1 km long, the error at the end would be 100 × 5 cm or 5 m which is quite unacceptable when 1 m is 1 mm on our map at 1/1000 scale. If the road is level and CD is also nearly so, the same error will be present in traversing back down the same slope from D to B and so it will not cause any misclosure in our traverse circuit ABCD. However, the position of CFD will be shown 5 m too far from the road. This error increases as the square of the slope.

34 SIMPLE PLANIMETRIC SURVEYING

At this stage it is worth examining the whole subject of errors, since we are going to have to be much clearer in our minds about them in traversing than in the relatively simple techniques of chain survey or triangulation where the geometrical figures are simple and easily and quickly checked. Two main points should be borne in mind: the first is the threefold nature of all errors in measurements taken for scientific purposes, and the second is the relationship between distance errors and angular errors which is important to surveyors and engineers. Measurement errors are of three kinds.

(i) *Accidental*: small observational errors in measuring each length or bearing (or any other quantity).
(ii) *Gross*: blunders or mistakes in which large misreadings occur.
(iii) *Systematic*: e.g. using a tape which is too long or too short (or a thermometer with a displaced scale).

The accidental errors are of the order of the precision of reading, that is to say about a centimetre on a tape and half a degree in a hand compass. They are equally likely to be positive and negative, and if the total quantity measured is the result of an accumulation of measurements, as in taping a 300 m length with a 30 m tape, the error at the end is likely to be equal to the accidental error of each measurement multiplied by the square root of the number of measurements. In this case the probable error at the end would be about $\pm 1 \times \sqrt{10} = \pm 3$ cm in ideal conditions, but what with errors caused by incorrect tension in the tape and in step chaining it might be nearer ± 5 cm in each bay even in normal conditions and so ± 15 cm at the end. If a fresh bearing is taken with the compass along each leg then there is no accumulation of error, but if angles are used then the bearing error at the end with ten legs might be $\frac{1}{2}° \times \sqrt{10} = 1\frac{1}{2}°$.

The errors which cause most trouble in cumulative measurements are the systematic ones because what may seem quite a small error in one measurement can soon grow to unacceptable dimensions if repeated. Examples are constantly stretching the tape more or less than the maker intended and thus increasing or decreasing its nominal or calibrated length throughout the whole distance, not correcting sloping legs, not having the tape straight (which also always makes the measured distance longer), the existence of an error in the tape itself (e.g. through a wrong repair after a break), and so on. As we saw earlier, one danger of systematic errors is that they can only be discovered by a completely independent check; for example, if all length measurements are wrong by 1 per cent then in all our three surveys the figures and traverses could close satisfactorily and the map would show the correct shapes but it would be on the wrong scale by 1 per cent. This error could only be discovered by a completely independent survey using a different tape. Gross errors are easily detected when the survey framework is simple, but if the error is not very large (say 1 m) in any one traverse leg it can be swamped by the accumulated accidental errors in a long traverse.

The other point which concerns surveyors is the relationship between errors in distance and errors in length. As most people know, or can discover by using a protractor and a scale, an angle of 1° corresponds to, or subtends, a distance of roughly 1 part in 50 of the length of the two arms of the angle and this direct relationship holds good for all small angles up to about 10°. Thus, since errors in angles or bearings should be well under 10°, we can say that an error of half a degree in a bearing from one station to another causes an error at right angles to the bearing of one hundredth of the distance between them. Clearly, therefore, if distance measurements are to be consistent in accuracy with those of bearings or angles, we really only need to measure distances to about the same accuracy of 1/100, which in compass and tape survey would be about 30 cm for each 30 m tape length. In fact, we can do a good deal better than this even on sloping ground, and we shall take advantage of this fact in adjusting our framework. However, it is also worth noting that in a rough survey, e.g. a reconnaissance of an area or for detail mapping as described on p. 31, we can do quite well by walking along the paths, counting paces, and taking a series of bearings along each straight section of the path since careful pacing can be accurate to about 1 part in 50 and bearings taken in this way without back bearings or without using ranging rods set up over our stations, or marking these with pegs, are unlikely to be much better than a degree in accuracy which is also 1 part in 50. In fact a reconnaissance survey made in this way might be worth while as a check on a gross error such as miscounting the tape lengths along a traverse leg.

Returning to our traversing, we assume that the scientist has carefully measured the lengths and bearings of all the legs along the four main traverses AB, AC, CD, and DC. How does he plot them? AB should be plotted first, and the two ends of this are marked on the map with circles, because since they are unchecked they are still provisional. Next AC should be plotted starting from A and taking care with the

36 SIMPLE PLANIMETRIC SURVEYING

bearings because plotting fairly short legs can introduce an error as large as the observation if we are not careful; a sharp hard pencil should be used. In order to preserve the bearings all over the sheet a grid will have to be drawn on it. This is done with the ruler by drawing as large a square as possible using the arc method and then extending this to cover the sheet; the sides are then divided into units of (say), 10 cm so that wherever the protractor is used it can be accurately aligned parallel to the sides of the grid squares. Traverse BD is then plotted on the map from B, and traverse CD is plotted on a separate sheet of tracing paper and placed over the map so that C falls over the position plotted from AC (see Fig. 2.7). We shall find (unless we are very

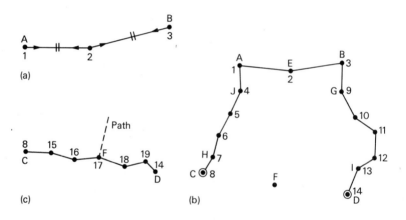

Fig. 2.7 Plotting the traverses: (a) traverse AB; (b) traverses AC and BD; (c) traverse CD (on tracing paper). Each traverse station has a unique number.

lucky indeed) that D on the tracing paper falls at D′ on the map as shown in Fig. 2.8(a) with an error l in the length and b in the bearing. (These are exaggerated in the diagram.) We have seen that with our equipment lengths should be more accurate than bearings, and this *closing error* is in fact made up of small errors in length in all three traces and larger errors in the bearings (Fig. 2.8(b)). Provided that

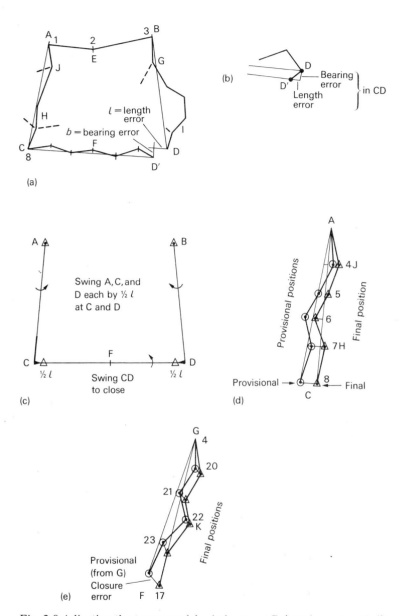

Fig. 2.8 Adjusting the traverses: (a) misclosure at D (much exaggerated); (b) length and bearing errors; (c) adjusting the three traverses to close; (d) adjusting traverse AC (bearing only); (e) adjusting traverse GF (bearing and distance error).

l and b are of the order to be expected over distances of about 1 km, i.e. a few millimetres on the map, we can close the figure best by assuming that all the errors are in the bearings and that the lengths are errorless. We swing AC and BD towards each other by $\frac{1}{2}l$ (Fig. 2.8 (c)) and swing CD at D northwards by b. CD on the tracing paper will then fit exactly over the shortened CD on the map but will have not altered at all the lengths of the three traverses, only their bearings. Having achieved this closure we mark the five points ABCD with triangles as confirmed and adjust the intervening traverse stations along each as shown in Fig. 2.8(d). We are now in a position to observe (or if already observed to plot) the remaining traverses, accepting as fixed the confirmed positions of E, F, G, H, J, and I. It will be best to plot GF first because it is the shortest, and then use K as an end point for plotting HK and IK separately. Misclosures are adjusted as shown in Fig. 2.8(e).

Suppose, however, that the circuit ABCD does not close by some 2 or 3 cm on the map which is a good deal more than we should expect. Since the bearings have been checked by back bearings, and there are relatively few of these, the gross error is more likely to be in a measured length. If it is north or south then it will be in AC or BD, and if it is east or west then it will be in CD. Before going out to remeasure these distances it will pay to observe and plot GF and see where F falls in relation to its position on CD as laid down either from C or from D. If it agrees with the first then it is likely that the error lies in the stretch GD. There are of course various possibilities, and while searching for the error we should endeavour to build up those parts of the framework which close well and by this means isolate the error in two or three traverses, remembering that if it is a gross length error it is likely to be in tens of metres and if it is a gross bearing error deduced from an angular error of $10°$ there will be a swing of this amount over the stretch on one side of the error. Another possible source of bearing errors is sighting to the wrong station or to a ranging rod not placed correctly over the peg from which the next bearing is taken, but this should show up in a discrepancy between forward and back bearings. If repeated discrepancies turn up between forward and back bearings and they cannot be eliminated by checking, then there must be magnetic anomalies in the area and the observed angles must be used instead of bearings. If this occurs no leg should be less than 50 m long. In each circuit the total number n of internal angles should be added together; the sum should be equal

SIMPLE PLANIMETRIC SURVEYING 39

to $(n - 2) \times 180°$, e.g. a ten-sided figure should add up to $8 \times 180° = 1440°$. A closing error of $\frac{1}{2}°$ times the square root of the number of angles may be expected, say 1° or 2° and this is distributed equally through the circuit. Plotting will have to be very carefully done because when legs are used only a few centimetres long on the map the errors can be as large as the observing ones.

The foregoing shows that traversing is a laborious technique, with many sources of error which may be difficult to identify; if even a part of the area is clear enough for triangulation this is to be preferred. If some distant object is visible from a number of traverse stations, it is also a useful check to take bearings to it from all of them since plotting these will give an additional check on the traverse.

We have now established a framework. What about the detail survey? If the area is really densely forested then we cannot see anything that lies off the tracks, but assuming that we can see short distances through the trees to the detail we want to map, such as streams or different species of tree, we do this by more traversing but, as in the previous sites, to a lower standard than the surveys for the framework because the accumulation of error is kept under control by the framework. These traverses can have much shorter legs; we need not put in pegs, the distances can be measured with a rope of known length stretched between two observers, and the bearings can even be taken, not to a visible object, but to the sound of a whistle or shout from the man pulling on the front end of the rope who is invisible among the trees. This rope and sound traverse is the lowest form of survey known to man, but it was used for thousands of miles of detail survey in West Africa before the days of aerial survey and produced surprisingly accurate results when it was controlled by traversing of a higher order carried out with a theodolite and steel tape. We can also use this method for surveying something which may itself be rather indefinite, such as the boundary between two different species of tree. It is a cardinal rule that detail traverses must start from one traverse station of the main framework, or from a point fixed at distances measured along the leg between two of them, and must finish on a similar point the position of which is known with plottable accuracy on the map. Pacing can even be used for distance measurement in detail survey if the ground is reasonably flat; if we can achieve an accuracy of about 1 part in 50 in length and bearing then if necessary we can run a traverse up 10 cm long on the map (100 m on the ground) which should misclose by about 2 mm and not be inaccurate to more than this in its middle after adjustment. Such traverses should be as straight as possible; obviously the shorter we can keep them between checks the more accurate they will be.

At the beginning of this chapter we described how to set out a grid on a flat area; this is also quite often required in forests either for sampling purposes or for establishing lines of levels for contouring. It is often impossible to do this by air survey because only the canopy of the forest can be seen and the shape of this may be appreciably different from the shape of the ground underneath. Samples taken along paths or streams may be biased because the paths have been chosen to follow the drier ground while the streams obviously follow the wettest areas. To set out a grid the perimeter traverse must first be completed, closed, and plotted on the map – unless of course the boundary of the area has already been surveyed and mapped. When the spacing and direction of the parallel cut lines have been decided, starting points are chosen along the perimeter on the map and their positions are fixed on the ground either by map reading or by measurement along the leg of the perimeter traverse on which they lie; in both cases the position should be fixed by at least two measurements from existing detail or traverse stations. The direction of each line is set out by protractor on the map and the corresponding compass bearing (allowing for local deviation of the compass) is worked out. The scientist then starts each cutting party along a line until they have cut at least 100 m. Poles are put in at the starting peg and at the end of the cut and are aligned along the required bearing; they must be intervisible. All the cutting party then has to do is to sight back along the two poles and keep them in line until just before the starting pole is hidden by intervening ground; a third pole is then put in in line with the other two. Repetition of this technique will take the party up to about 0.5 km, but the line is then likely to start to deviate from the bearing and the scientist must visit each party regularly to check on the bearing of their line and correct it if necessary. Eventually it will arrive either at the other side of the perimeter or at an internal path which he has already mapped. The position of this cut should be surveyed by measurement from the nearest features or traverse stations along the perimeter or internal path so that the actual (as opposed to the theoretical) positions of the cut lines can be marked accurately on the map. The samples or levels etc. can then be set out and plotted by measuring the distances along the lines and transferring them to the map.

3. HEIGHT MEASUREMENT

Introduction

So far we have only considered planimetric mapping, i.e. without spot heights or contours; in this section we describe the three principal techniques of height measurement ('heighting') in their simplest forms, suitable for adding heights and contours to the maps of the three areas already discussed. The three techniques are spirit levelling, trigonometrical heighting, and aneroid barometer heighting. Which technique to choose in particular circumstances depends both on the nature of the terrain and on the use to which the map is to be put. Both these criteria tend to fall into two categories. In the first category the area is flat, and in such circumstances the heights are usually required for the design of open-water channels or for the study of hydrology and marsh ecology for which very accurate heights are necessary. In the second category the ground is hilly and contours are required to show the shape of the ground for ecological or soil surveys, for planning building or road layouts, for studying soil erosion and its prevention, or for working out how to transport logs out of a forest. In such conditions accurate heights are more difficult to obtain and less important, and an accuracy to a foot or even a metre may be quite adequate; it is the slopes that matter rather than the actual heights. This chapter also describes two techniques of three-dimensional mapping, by plane table and from ground photography.

Spirit levelling

Where heights to the nearest centimetre or so are required the only practicable technique is that of *spirit levelling* which is very simple in theory but tedious in practice. In its simplest form, as shown in Figs. 3.1(a) and 3.1(b) the spirit level consists of a telescope, which includes a diaphragm with cross hairs or a graticule engraved on glass, and a glass tube or vial fixed parallel to the telescope. This tube is not quite filled with spirit so that it includes a bubble, and it is slightly curved in the vertical plane so that the bubble floats to the highest part of the curve; two marks are engraved at equal distances from its

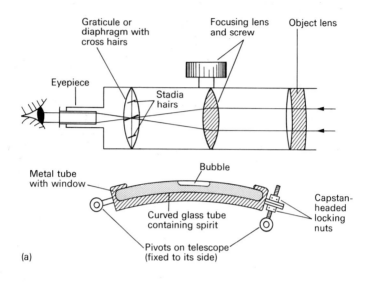
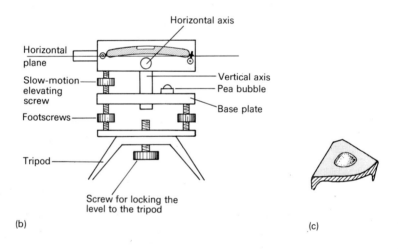

Fig. 3.1 The spirit level: (a) elements of the spirit level; (b) spirit level supports; (c) crow's feet for staff support.

ends so that when the bubble is symmetrical about these the ends of the tube are on the same horizontal line. The sensitivity of the bubble is directly related to the radius of curvature of the tube. This is fixed to the side of the telescope with a pivot at one end and a ring fitting over a vertical threaded bolt on which are two lock nuts at the other. The lock nuts are usually round and have holes so that they can be turned with a tommy bar and not a spanner; they are called capstan-headed nuts. These must be adjusted so that the axis of the level tube is exactly parallel to that of the telescope as defined by the centres of its lenses and the intersection of the cross hairs. The telescope has two axes of rotation. The first is vertical and allows for rotation so that it can be pointed in any horizontal direction. The second is horizontal and perpendicular to the telescope axis; it allows for limited rotation in the vertical plane so that the telescope can be made exactly horizontal; for this purpose a slow-motion elevating screw is attached near the eye end.* After pointing at a desired object the horizontal movement can be stopped by a clamp and the precise direction of pointing adjusted, while looking through the telescope, by a slow-motion screw. The base plate is attached to the tripod either by a ball joint and clamp, or by three foot screws so that it can be roughly levelled using a small coarse circular pea bubble fixed to its upper surface. When this has been done the more sensitive telescope bubble should stay in its run, i.e. not up against either end of the tube, whatever the direction in which the telescope is pointed. The tripod legs must be well trodden into the ground and, assuming that the telescope and the level tube are truly parallel, when the bubble is brought to the centre of its run by the slow-motion screw any object seen through the telescope in line with the horizontal cross hair of the graticule will be on the same horizontal plane as this. In most levels the bubble can be seen by a mirror at the eyepiece end of the telescope.

In order to measure the relative heights of different parts of the ground we use a vertical staff graduated in feet or metres with zero at the bottom. Various forms of graduation and numbering are used and the observer should make himself familiar with that on the staff with which he is supplied before starting to use it; it is very easy to make misreadings unless this is done. If this staff is placed on a series of points in turn then the differences between the staff readings at them will be exactly equal but opposite in sign to the differences in their

---

*In the *dumpy level* this is done by the foot screws only, but is less accurate because the adjustment is coarser and it may change the height of the instrument.

heights as shown in Fig. 3.2. If we set up the level in the middle of the first area then the height of any other point can be measured by setting the staff on it and taking the reading where the cross hair cuts it. Comparing these readings will give *relative* heights, and in order to turn them into *absolute* heights we designate one of the framework

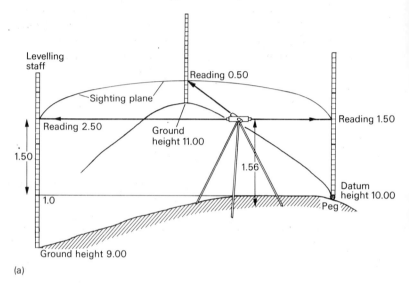

(a)

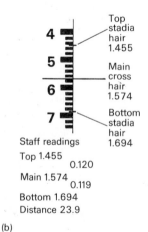

(b)

Fig. 3.2 The theory of spirit levelling.

pegs as a datum and allot it a sufficiently large value (in this case say 10.000 m) so that none of the heights in the area will be negative. Normally the heights observed for points on which the staff is set are more accurate than the height of the ground under the instrument itself, but we can determine this either by setting the staff against its horizontal axis or in some later models by a graduated plumbing rod which extends to touch the ground vertically underneath. In observing the staff at different distances it will be necessary to change the distant telescope focus (but not that of the eyepiece) and this should be checked by small movements of the head which should not, when it is correctly focused, produce any movement of the staff image relative to the image of the cross hairs. The staff appears inverted through the telescopes of most levels (as of most survey instruments) and the numbers on it are painted upside down so that they are easier to read. If an erect-reading instrument is used the staff should also be erect-reading. It will be noticed that the graticule includes two short lines crossing the vertical line or hair at equal distances above and below the main horizontal cross hair. These are *stadia hairs* and they can be used for measuring distances from the staff and also as a check since they are equally spaced and subtend an angle of exactly 1/100 or about 0.5°. Thus the distance between the instrument and the staff is 100 times the difference between the two stadia hair readings and 200 times the difference between one stadia hair and the centre hair. The readings on them also provide a check on each other and on the main reading since this should be exactly half way between them. For simplicity these readings have not been included in the examples, but clearly this is an easy and rapid way of measuring distances which can be used for fixing the positions of levelled points.

The process described, of setting up the level on one point in the middle and observing from this one set-up to a staff placed in turn on all the others, provides no check and could only be done on a relatively small site. Therefore we would normally follow the same principle as in planimetric survey of putting in a framework of more accurate and reliable points, permanently marked by pegs with smooth or flat or domed tops, before starting to fill in detailed heights. This is done by placing four corner pegs round the perimeter. We then put the staff on each of two corner pegs in turn with the level set up roughly midway between them; the staff is observed from each set-up of this kind on each of the two pegs in turn. In this way we observe the successive differences between all four pegs and complete the circuit by putting the staff on the first one at the end. With a staff graduated in metres and centimetres, we should then obtain readings along the lines shown in Table 3.1. It will be noticed that we write the backsights and foresights, and also the rises and falls, in separate columns; this makes it easier to add them up at the end which is an invaluable check on our arithmetic, for as will be seen the difference between columns 3 and 5 is equal to the difference between columns 6 and 7, and also to the *closing error* on A.

In fact, a misclosure of 5 cm is rather large round a circuit of only about 400 m and with only four set-ups. What has happened is that the level tube was not parallel to the telescope axis, and so instead of the latter describing a horizontal plane as it swept round it described a flat cone (Fig. 3.3(a)) and the reading on the staff was in error by an amount which depended on its distance from the level. We should now do what we should really have done at the start: check the adjustment of our instrument. This is done by what is called the two-peg test which is shown in Fig. 3.3(b). To reduce the error we tried to set up midway between successive staff stations, and to find it we do exactly the opposite and set up alternately very near to each of two pegs in turn when carrying out the test. If we do this between A and B we might obtain the following results:

(First set-up near A): to A 1.61, to B 1.87, fall 0.26
(Second set-up near B): to A 1.38 to B 1.56 fall 0.18
Mean is 0.22, difference 0.08.

The error at each set-up was half the difference or 0.04, and in the second set-up it will all be in the long sight to A. We therefore turn the level back to A and by trial and error adjust the end of the bubble tube so that with the bubble in the middle of its run the reading to A is now 1.34. The reading to the much closer point B should be unchanged at 1.56 and the difference will now be correct at 0.22, so the level is in adjustment; however, we can check this by repeating the observations from the set-up near A. If we now repeat the circuit we should obtain a very small misclosure.

This main adjustment of the level is the most important way of eliminating minor accidental errors in spirit levelling, but if the lines are longer (for example along the paths in the third site assuming that this was fairly flat) then other sources of cumulative or systematic error must be watched. We must check that the tripod legs of the level are firmly embedded at each set-up or it may change height between

TABLE 3.1

Set up between	Backsight to	Value of backsight	Foresight to	Value of foresight	Rise	Fall	Reduced level	Peg
A and B	A	1.53	B	1.76		0.23	10.00	A
							9.77	B
B and C	B	1.98	C	1.07	0.91		10.68	C
C and D	C	1.02	D	1.38		0.36	10.32	D
D and A	D	1.56	A	1.83		0.27	10.05	A
		6.09		6.04	0.91	0.86	+0.05	
			+0.05		+0.05			

48 HEIGHT MEASUREMENT

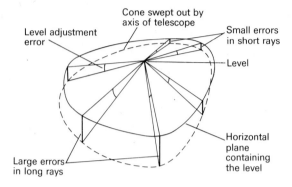

(a)

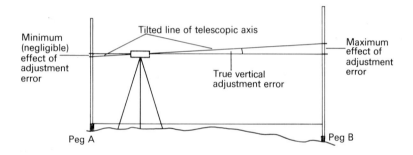

(b)

Fig. 3.3 Collimation error: (a) the effects of maladjusted levelling from one set-up; (b) the two-peg level test (first set-up near A).

the backsight and the foresight. We must also check that the staff is placed on firm ground or on a peg or heavy base plate or 'crow's foot' which is well trodden in (Fig. 3.1(c)), otherwise it may also change height between the foresight to it from one set-up and the backsight to it from the next one as it and the level leapfrog each other along the line. The staff must be held upright, for which a small hand bubble can be used, or waved back and forth with the observer taking the minimum reading which will clearly be when it is at right angles to the line of sight and is therefore upright. Its foot must also be kept clean of earth or stones. When we have levelled over the perimeter circuit and closed it satisfactorily and also levelled over any internal lines required to establish a good framework of reliable heighted pegs (or *benchmarks* as they are called), detail heights can now be observed either by single set-ups starting with the staff on a benchmark or by running lines of levels from one benchmark to another or by closing a circuit back on the same benchmark. As in fixing planimetric detail points, the positions are obtained by a suitable combination of measured distances and bearings or angles; the distances can be measured by tape or by using the readings on two stadia hairs. The sights are booked as shown in Table 3.2 where a large number of intermediate heights have been observed from one set-up before using the last of these (g) as a *change point* from which to leapfrog the level on to a further set-up from which a number of intermediate spot heights can again be observed before closing on the second benchmark at B. With intermediate heights special care must be taken not to make gross errors or misreadings since there is no check on these except, as in the check on planimetric detail points, the observed shape of the ground or feature in between them which should agree with the recorded difference between their values and positions. The plot of the results is shown in Fig. 3.4.

In this part of the book we have been considering survey procedures — including the production of heights — on relatively small areas. However, since levelling is much the same whatever the size of the area it is convenient to include at this stage the changes in technique required for levelling large flat areas. The first change is to take much greater care to avoid mistakes, because with much longer circuits the time wasted in failing to close a circuit is obviously much greater than when it has only taken about an hour. The second change is that since levelling is a simple and tediously repetitive procedure it is better carried out by less rather than well educated personnel, with the scientist or a senior surveyor acting as a supervisor. In large areas the framework levelling may take several days — or even weeks — and a mistake at one station may require the re-levelling of a circuit which may take several days to complete. As in all surveying, mistakes are best avoided by constant checking at the early stages of the work; in the case of levelling checking is best done after the staff has been read and before it or the instrument is moved. The two commonest mistakes in levelling are (i) misreading of the staff by a major unit (1 ft or 0.1 m) and (ii) failure to level the bubble at the time of reading.

TABLE 3.2
Record of the level survey

Object	Bearing	Distance	Backsight	Foresight	Intermediate	Rise from the last backsight	Fall from the last backsight	Provisional reduced level	Correction	Final reduced level
A (BM)	310	531	1.52					10.00	(datum)	10.00A
B	64	—	—		—					
a	358	23.2			1.65		0.13	9.87		
b (bottom)	31	32.5			1.78		0.26	9.74		
c	85	25.5			1.49	0.03		10.03		
d	141	28.6			1.27	0.25		10.25		
e	208	28.0			1.16	0.36		10.36		
Instrument	00	00			1.56		0.04	9.96		
f	279	34.5	1.48		1.40	0.12		10.12	+0.01	10.13
g	65	25.0		1.76			0.24	9.76	+0.01	9.77
B (BM)	54	57.5		1.72			0.24	9.52	+0.02	9.54B
A	79.5	—					—	—		
Instrument	00	00			1.49		0.01	9.77	+0.01	
		checks	3.00	3.48			0.48	−0.48		

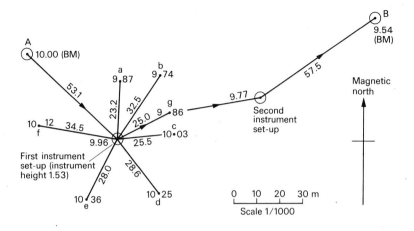

Fig. 3.4 Plot of level survey.

The best way to eliminate (i) is to read on all three hairs of the telescope; it only takes a few seconds to do this and an immediate check on the centre reading – the one that matters – is that it should be midway (within a millimetre or so) between the other two. The upper and lower readings should be recorded in smaller numbers than the centre reading. Failure to check the bubble is more difficult to prevent, but if this is done both before and after reading the centre hair as it should be, then the reading can be ticked to prove that it has been done. Before staff or instrument are moved the leveller should check that the three readings are consistent and that the centre reading has been ticked. An advantage of this procedure is that if the circuit still miscloses a scrutiny of the record book may disclose where the error probably occurred, and if readings were taken on to a peg or other hard identifiable object near by it may only be necessary to repeat a few set-ups near that point and not have to re-level the whole circuit. In addition to these precautions it may also be worth while to double level the whole circuit and break it down into relatively short sections between pegs or benchmarks. The second levelling between these, which is done in the opposite direction, should then confirm the first levelling, and re-levelling will only be necessary in those sections where the two runs of levels disagree. Having taught himself how to level successfully, the scientist will be in a position to consider the second change required for levelling large areas, i.e. the employment of semi-skilled assistants or of a contractor.

Spirit levelling along a line is in some ways the simplest of all survey techniques; for this very reason an intelligent and well-qualified scientist might find, after he has mastered the technique, that carrying out any major detailed levelling himself would not only be very tedious but would also be a waste of his valuable and presumably expensive time. If a good deal of levelling is required, as it is in any area being considered for irrigation, then it must be done either on contract or by engaging local levellers to work directly under his supervision. In each case it is worth stating the basic principles of such an operation. The work must be planned and laid out in a proper manner with a framework of higher accuracy levels for which the person in charge should be responsible himself, and where relatively low-grade assistants are used for the detailed levelling only he should know the results of it. At its simplest this framework may be a circuit round the area, but in more extensive projects it will also include a number of cross lines. These should be well marked by permanent benchmarks, especially at intersections, and pegs should be put in along them at the ends of the detail lines along which spot heights are to be measured at intervals. The instruments and staves should be checked and adjusted where necessary; in the tropics an umbrella – and a man to hold it – will be required to shade the instrument. Up to four parties, each comprising one leveller and a staff holder, can be supervised by one experienced professional or senior technician surveyor, or by the scientist himself if he has only a small amount of other work to do.

It takes several months to train levellers of a relatively low educational standard, and therefore these would normally be recruited from those who have done similar work before or be borrowed from the local survey department. They may of course be undergraduates studying engineering, geography, or similar subjects who have already been taught surveying. The principal requirements, apart from education up to at least one or two years of secondary school, are honesty, reliability, and neatness in keeping records. The levellers should be issued with numbered books and told to keep all their records in these; all the books must be accounted for. This is to prevent any possibility of bookings being copied for greater neatness or to *fudge* or adjust them so as to appear to close onto framework benchmarks in cases where a mistake has been made. This can also be prevented by not allowing the assistants to know the values of the benchmarks on which

HEIGHT MEASUREMENT 53

they start and close their level lines; however this procedure should only be applied to poorly educated levellers and not to undergraduates who should be encouraged to take an interest and pride in their work and in their own integrity. Frequent unannounced visits should be made to the levellers in the field to see that they are following the proper procedures, and the instruments should be checked every few days to see that they are properly adjusted. The books should be checked every evening and scrutinized by the supervisor who should be looking out for erasures – which are strictly forbidden – or for too regular writing which may indicate copying in the tent or office from a less neat record made in the field – an even worse crime. The sums of rises and falls will indicate the total rise or fall along each line between the benchmarks or the pegs at the end of each detail line, and this can be compared with the values of these kept by the supervisor; if they are satisfactory the reduced levels can be computed and inked in. This work must not be allowed to fall behind, and a record should be kept of the number of lines levelled by each individual which have to re-levelled; where this proportion is too high he will have to be replaced. Where four levellers are employed it will be worth while to engage a computer to check the reductions of their work; he and the levellers should be supplied with hand calculators. To sum up, detailed levelling of large flat areas is a tedious task in which too much education may be a disadvantage, but even the less well educated can work out how to record observations which they have not actually made or how to adjust those which have been wrongly made if they know what the answers, in the form of line closures, ought to be. This kind of mass production of survey data will only be done properly by semi-skilled staff if they are correctly instructed and closely supervised so that these dangers are guarded against. They must not be left to carry out the work without supervision.

It is clear that so far we have only produced a series of spot heights and not contours; in levelling flat areas the contours are usually interpolated afterwards between the spot heights. If the area is very flat and open the spot heights can be set out in a regular grid and assistants can be used. Where the form of the slopes can be perceived by eye the spot heights should be sited on the tops or bottoms of rises and dips and at changes of slope so that the positions of the contours can be interpolated directly along the uniform slope between adjacent spot heights. A better qualified leveller is required for this sort of work and would find it interesting. The method of booking and the

54 HEIGHT MEASUREMENTS

plotted result of a small part of a detailed survey of this kind are shown in Table 3.2 and Fig. 3.4 respectively, and the following points should be noted:

(i) All the data about one staff set-up i.e. foresights and backsights to it, its rise or fall relative to the previous point, and its reduced level, are plotted on one line.

(ii) All three stadia hairs will have been read (but are not shown), to measure the distances and as a check against misreading.

(iii) Separate columns are used for backsights, foresights, and intermediate sights so that the latter do not interfere with the summation checks which have to be applied as described on p. 46. In the example shown in Table 3.2 and Fig. 3.4 g to B with only two set-ups; each intermediate shot is on its own. None have been shown from the second set-up, but of course normally there would be some from this also.

(iv) The instrument height is treated as an intermediate shot.

(v) There is a small misclosure of 2 cm on B which we ignore except as confirmation that the levelling was done correctly.

(vi) In the plotting the actual position of each spot height is shown by its decimal point; this is a neat solution to the problem of where to write its value.

In Fig. 3.5 a series of spot heights set out in the form of a grid is shown with 10 cm contours interpolated between them. For greater accuracy all the distances between adjacent spot heights should be

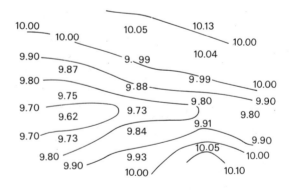

Fig. 3.5 Interpolation of contours in a flat area.

HEIGHT MEASUREMENT 55

divided in the exact proportions of the differences between the spot heights and the contour to decide where it cuts the line between them. Elaborate computer programmes are devised to take account not only of the adjacent spot heights but also of others further off which may indicate the curvature of the ground; however, this is only necessary in large and expensive projects.

Other techniques

Trigonometrical heighting

It is clear that levelling would be both inappropriate and difficult for the second site described in Chapter 2, in view of the steep slopes and the short sights which would require many set-ups of the spirit level. The level cannot be set-up much higher than the observer's eye (about 1.5 m above the ground), and so if the point being levelled to is at a greater height than this above the ground on which the level is set up it will not be possible to intersect the staff with a horizontal sight; on a slope of 1 in 10 this gives a sight length of only 15 m. One possibility is to use a telescope which can be tilted and to measure the angle of tilt. This is called *tacheometric levelling*, but it requires the use of a theodolite which is too complicated and expensive for use in simple surveying. What in fact we do is to use a simpler instrument, called an *Abney level* or *clinometer*, in which we observe vertical angles against a small spirit level; since the horizontal distance between two points is known from our planimetric map we can then calculate by simple trigonometry the difference in height between them. When this instrument (shown in Fig. 3.6) is held steady against a staff or a tall stick, we can observe vertical angles with an accuracy to about $10'$ or, as we know from the previous comparison between angles and distances, about $1/(50 \times 6)$ or $1/300$ of the horizontal distance. The ratio between the vertical and horizontal distances in a triangle of this shape is the tangent of the vertical angle; this can be obtained either from trigonometrical tables or nowadays from a calculator costing less than £20 which can also be used to multiply the two together and so obtain the difference in height. The horizontal distance can be scaled from the map (if the two points have already been fixed) or obtained by any of the methods described in Chapter 2. Allowance must be made for the height of the observer's eye above the ground on which he stands, either by including it in the equation to derive the true height difference, or more easily by observing to

56 HEIGHT MEASUREMENT

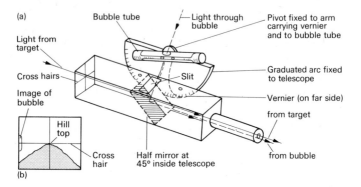

Fig. 3.6 (a) The Abney level. (b) View through the Abney level.

a point at an equal height above the ground level at the target end of the line such as the head of an assistant or a rag tied round a ranging rod at this level. As in levelling the heights of the framework pegs should be obtained first, using one as a datum; detail points can then be heighted either by taking shots from the framework points into them, or by occupying the detail points and taking shots out to the framework pegs, of which two should be used to obtain a check.

In this technique it pays to plot the results as we go along so that the contours can be sketched in by eye between the heighted points; the points will not be set out in a grid but will be selected to give heights between which slopes are regular, and of course on the tops of hills, at the bottoms of valleys, and on saddles and shoulders. The sort of result which is achieved is shown in Fig. 3.7 where it can be seen that several contours can sometimes be interpolated between two adjacent spot heights where the slopes are steep. With an angular error of 10′ or 1/300 we can achieve height accuracies of the order of 0.5 m by keeping sights down to about 100 m (4 cm at 1/2500 scale on the map), and if we can make our framework heights agree with this sort of accuracy we can plot 2.5 m contours with confidence; on a 1/2500 scale map and a slope of 1 in 10 these will be 25 m or 1 cm apart.

Barometric heighting

The other simple heighting technique appropriate in these conditions is to measure the differences in air pressure at different points with an *aneroid barometer*. This is less accurate since even the largest and most

HEIGHT MEASUREMENT 57

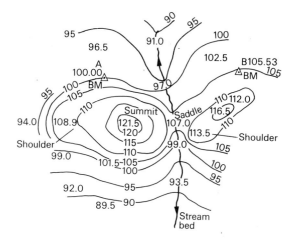

Fig. 3.7 Heighting and contouring in a hilly area.

precise aneroids will only read to the nearest foot or 0.5 m, and the heights will not be accurate to better than 1 m even under good stable atmospheric conditions. The aneroid barometer (Fig. 3.8) consists essentially of a series of flat cylindrical evacuated capsules with diaphragm sides which move in and out by small amounts as the air pressure increases or decreases. These small movements are enlarged by a series of linkages to turn a pointer over a circular scale which can be graduated in height values of metres or feet, or in pressure values of inches of mercury or millibars. The former are the most convenient for our use at this stage since we shall be using the directly observed differences and not calculating them from the actual observations of air pressure which have to be corrected for temperature etc. As long as frequent visits

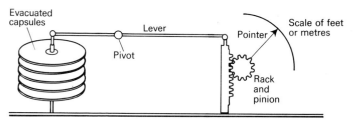

Fig. 3.8 Simplified design of an aneroid barometer

are made to framework points of known heights to check on changes in the air pressure during the day no further corrections, apart from a small closing correction, are required. But if this is not possible then it will pay to have a second observer who remains stationary at a point of known height and observes a second barometer there to record at frequent timed intervals all the changes in pressure which occur either because of unsettled weather or by the heating of the ground in settled weather, especially in the topics where the daily variation or *diurnal wave* is quite large. The pressure observed by the 'roving' barometer is then corrected by interpolating between the times when these observations were taken. It should be noticed that while height differences observed by vertical angles require the positions or horizontal distances to be known, those observed by levelling or barometer are given directly, although of course a height without a position is not likely to be of much value. However, the point is that in levelling or barometric heighting the positioning and the heighting may be two quite different processes, and this makes them particularly suitable for use with air photographs since all that is required is to identify each heighted point on a photograph and leave the positioning of it to be done at a later date, possibly by someone else using special techniques in an office. Barometer heighting is also particularly suitable in work at very small scales when one's position can be fixed by map reading or by some form of navigation such as traversing by car (or sledge or camel) and compass which is obviously far too rapid for any other way of carrying heights.

Plane tabling

Three different techniques have now been described for both planimetric and height or contour surveys in which the simplest of instruments and a minimum of computation was used. A common feature of all these techniques was the measurement and recording in a book of distances, bearings or angles, staff readings, or pressure differences as the raw material of the survey; the checking of these data and their conversion to positions and heights on the drawing board were separate operations even if at times they were carried out in the field and not in the office. This section describes a fundamentally different type of survey in which the drawing board itself becomes, in the form of the *plane table*, the principal instrument, and is used to record the angles and distances of the planimetric survey so that separate recording

HEIGHT MEASUREMENT 59

is unnecessary. While this has the advantages of saving time, avoiding copying errors, and fixing the framework and detail points and interpolating detail and contours immediately in the field, it has the disadvantage that, since observing and plotting are combined in one operation, there is no separate record of the basic data which can be checked calmly and independently in the office either by the surveyor himself or by a supervisor. Because of this, and because heights can also be derived directly in the field, both detail and contours can be mapped simultaneously, but as a result plane tabling requires (and of course also teaches) a much higher degree of integrity, skill, and understanding of survey principles than any of the separate technqiues already described, which is why it has been left till the end of this chapter. Moreover, when mistakes have occurred and things do not come out as they should — which happens to every plane tabler sooner or later — the problem has to be sorted out in the field, perhaps on some cold and windy hill top or in a village surrounded by small boys and plagued by flies, instead of in the peace and comfort of the office where the original data can be looked at calmly and analysed afresh to find the mistake. In spite of this, it is the most attractive and stimulating of all the survey instruments which the field scientist can use if the country is open enough, because one can see the map growing as one proceeds and because it is flexible; it is capable of quite high accuracy but it can also be used for rapid reconnaissance surveys to obtain an idea of the lie of the land as a preliminary to a more detailed and accurate survey by other techniques. Of the three sites so far considered, only the second (p. 22) is really suitable for plane tabling; the first is too small and on too large a scale, and the restricted visibility of the third removes the main advantage of the plane table of being able to see and record directly the shape of the ground and of the features on it and to measure angles and directions accurately over long distances.

The essential features of the plane table itself (Fig. 3.9) are a strong flat board rather less than a metre square and often rectangular, and a tripod to which it can be clamped by a vertical screw underneath it so that while it remains horizontal it can either be rotated freely about a vertical axis or clamped firmly in one position. Accessories include an *alidade* or sight rule, an *Indian clinometer* if heights are being surveyed, a trough or prismatic compass, a scale, some tracing paper, hard and soft pencils, and an india rubber. For large-scale work, where the size of the table is significant at the plotting scale, a *plumb bob* is

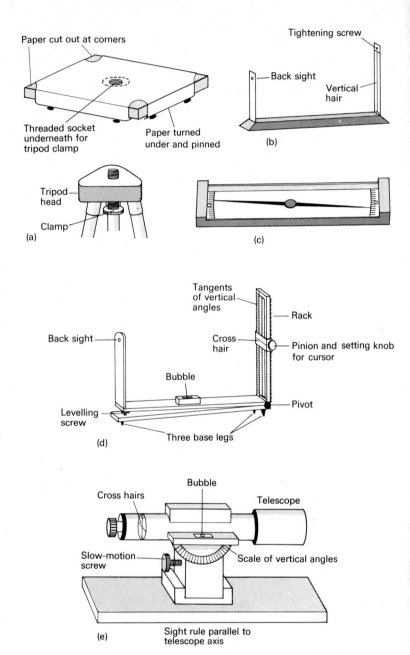

Fig. 3.9 Plane table and accessories: (a) the plane table; (b) sight rule or alidade; (c) trough compass; (d) Indian clinometer; (e) telescopic alidade.

also required, and for small-scale work binoculars may be required for identifying distant framework beacons or other objects. The alidade can be a simple sight rule with a folding vane at each end; that at the eye end has a vertical slit, and that at the object end has a fine vertical hair which can be tightened by a small set screw. An alternative is the *telescopic alidade* (Fig. 3.9(e)) which comprises a telescope, a bubble, and a graduated vertical scale of degrees. This is heavier, considerably more expensive, and more fragile than the simple alidade, and for most purposes it has few advantages to outweigh these drawbacks. In small-scale work the magnifying power of the telescope makes it easier to identify distant beacons on framework points, although this can usually be done without much difficulty using binoculars. In large-scale work with an assistant the stadia hairs in the telescope (similar to those already described for spirit levels (p. 45) make it possible to measure distances to a graduated staff held by the assistant and thus to fix points quickly by bearing and distance without having to use a tape. This technique can be very useful for establishing a large number of spot heights in a featureless undulating area. The actual mapping document is fixed to the top of the table. It can be a piece of good drawing paper cut to fit with enough to spare all round − cut away at the corners − for pinning underneath the edges of the table. To obtain a good flat surface without wrinkles it should be wetted first, then pinned, and allowed to dry, and this should be done before the grid is drawn on it. Nowadays it is more satisfactory to use some form of plastic cut to fit the top of the table and fastened to it by clips or drawing pins. This will not change shape or size and so a grid can be drawn on it at base before going overseas or to the field; it is waterproof, it gives a good flat surface, it can rapidly be changed for another map sheet, and finally it is transparent and so can easily be copied by the dyeline process. Useful materials are *Astrafoil* and *Permatrace* (see p. 283).

The essence of using a plane table to make a map is that it is *oriented* parallel to the ground so that all the rays drawn on it with the alidade from the map position of the occupied point must pass through the map positions of any points to which the alidade is pointed. The easiest way to orient the table correctly is to start from one end of a line joining two points whose positions are already on the table. These may be the ends of a base set out in approximately the same way as when starting the previous types of survey, or they may be two points of a more accurate framework which has been established by professional

62 HEIGHT MEASUREMENT

surveyors (or by the scientist himself) using techniques which will be described in later chapters. The plane table is set up horizontally by eye over the first point, the alidade is laid accurately along the line joining them on the map, and the table is then turned until the sights point to the other end of the base. The table is then clamped and should now be oriented correctly, i.e. parallel to the ground (Fig. 3.10(a)). If other fixed points (of a superior framework) are also

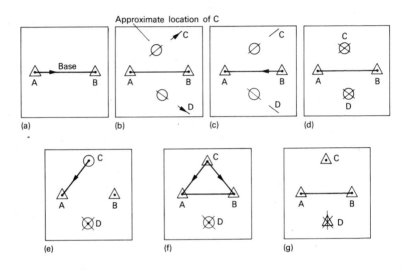

Fig. 3.10 Intersection of framework points: (a) table at A, sighted along AB to B and clamped; (b) table at A, rays drawn to C and D; (c) table at B, sighted to A and clamped; (d) table at B, rays drawn to C and D; (e) table at C, sighted along CA to A and clamped; (f) table at C, check sights to B and to A, positions confirmed; (g) check ray to D and if it passes through the cross there all four points are confirmed.

visible the alidade should also be pointed at them as a check; it should of course be exactly aligned along the rays to them from the occupied point. Rays can then be taken to any objects to be fixed exactly as in measuring their bearings by prismatic compass but using the sights of the alidade with its edge over the position of the occupied point. Each ray should be drawn with a hard sharp pencil in the area where the point seems to be, and also at the edge of the table; it should not be drawn through the occupied point. Identifications, with a sketch if necessary, should be written or drawn with a soft pencil at the edge of the table (Figs. 3.10(b) and 3.11).

The same procedure is now followed at the other end of the base line (Figs. 3.10(c) and 3.10(d)), and each of the points visible from both ends are shown on the table by a small cross round which a circle can be drawn and the point labelled with its name or number. This is all done in soft pencil as a provisional position. Each of these points should then be visited in turn, and the table should be oriented by sighting back on one of the base points (the further of the two) (Figs. 3.10(e) and 3.10(f)) and then checked by sighting on the other with the alidade laid against its position on the table and drawing the ray *back* through the provisional position of the station now being occupied. If it passes through the intersection of the previous two rays, then this position is proved, the circle can be erased and replaced by a small triangle, and the point itself accurately marked with a hard pencil. Any other points intersected from stations proved in this way and giving a series of rays all passing through one point can also be regarded as proved and be marked by point surrounded by a triangle in the same way (Fig. 3.10(g)) until the whole area to be surveyed has a series of such points widely distributed over it so that in surveying the detail the plane tabler can be sure of always being able to see at least three of them.

The reader will note that we have now measured, and recorded graphically but not in figures or digitally, the angle between each end of the base line and the ray or bearing to each of the other points. He should also notice that while the bearings measured with a compass and the derived angles are all equally accurate whatever the length of the ray, this is not the case in plane tabling because the accuracy of each ray depends on its length and on that of the ray used for orienting the table. Pencil lines can be drawn with an accuracy to about 0.3 mm, and over a 5 cm ray this means an angular accuracy to 1/150 or about 0.3° or about 20′. However, if the base line is formed by two points at opposite sides of the table and the full length of the alidade (say 30 cm) apart, the angular accuracy will be 0.3/300 or approximately 1/1000, which corresponds to about 3′. This means that a plane table survey can be to plottable accuracy (of about 0.5 mm) all over the table *provided* that its framework is established by using the longest possible rays first before starting work in the

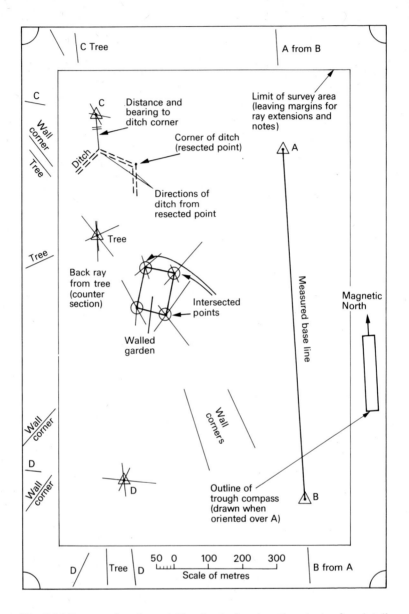

Fig. 3.11 Large-scale plane table sheet showing the start of a detail survey. The ray ends in the margin and the intersections etc. have not been rubbed out. This diagram shows a planimetric survey only. Points C and D and the tree were fixed before the detail survey was started.

middle; in other words it is even more important than before to work from the whole to the part and to establish (if it does not already exist from higher-order work) a framework right across the table before starting to put in detail points and draw detail and contours. If in open country this framework is built up by relatively short rays starting in the middle of the area, the time will come when a ray taken right across the table between two points far apart may miss the far point by an appreciable amount. This shows that the work is not accurate over the whole map, although it may be relatively accurate in small areas. To find this out at a late stage, when the map is nearly finished, causes the survey equivalent of weeping and gnashing of teeth.

Having established the main framework and recorded the direction of magnetic north on the table using the trough or prismatic compass, the plane tabler now proceeds to fix the detail points between which the actual lines representing the ground features, including contours if necessary, can be interpolated. There are four ways of fixing these: by *intersection*, by bearing and distance, by *countersection*, and by *resection*. The former has already been described; the procedure for establishing detail points is similar to that described in previous surveys. For bearing and distance the plane table is set up and oriented in the same way over a framework or detail point (provided the latter has been checked by a third ray), a ray is drawn from it to the point required, and the distance is measured by tape or pacing and set off along the ray by using a scale. In countersection (Fig.3.12) a similar forward ray is drawn, the plane tabler follows this until he reaches the point sighted to, and then sets up the plane table, orienting it by sighting back along the forward ray to the point he has just left. The

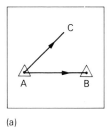 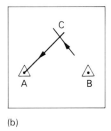

Fig. 3.12 Countersection: (a) at A, orient on B, draw ray to C; (b) at C, orient along ray back to A, draw ray back from B.

table is then clamped and a cross ray is drawn back from any other point visible from the new station. The position of the new point is at the intersection of the cross ray and the forward ray. If more than one other fixed point is visible from the new station the position can be checked by drawing the ray back from this also. Clearly, if the forward ray is short the orientation will not be very accurate; repeated application of this technique may lead to accumulated error, and before too many (say more than three) such successive points have been fixed the last one should be checked by some other method such as resection (Figs. 3.13(a) and 3.13 (b)).

The resection may well be full-scale, in which sights are taken to at least four framework points and both the orientation and the exact position of the occupied point are obtained. This technique, which is used a great deal by plane tablers in mapping some feature not readily visible such as a contour or soil boundary, is clearly more complicated because neither the position nor the orientation of the table is known precisely, although one of them is normally known approximately. In the case just cited it will be the position, but if the resection is started in the middle of an unmapped area then it will probably be the orientation which can be obtained using a magnetic compass, provided of course that the direction of magnetic north has already been recorded on the table when this was correctly oriented at some framework point as mentioned on p. 61. If an approximate position is known, but not the orientation, the next step is to lay the alidade from this position and to the furthest of the visible framework points and to rotate the table until the sights come onto the latter. The furthest point is used because, as we saw on p. 63, it gives the most accurate orientation for the table. If the orientation is known by compass measurements, the table is clamped so that this agrees with the direction of magnetic north previously drawn on it; the alidade is sighted on any one of the visible fixed points and a ray is drawn back from this map position in the general area where the occupied station is thought to lie. Similar rays are then drawn back from two of the other three or four fixed points visible from it (Fig. 3.13).

It is clear that while the *directions* to these points on the plane table will only be correct if its orientation is correct, the *angles* between them will always be correct. Thus the simplest method of resection is the same as that used with a compass, which is to place a piece of tracing paper on the table under the alidade and without moving the tracing paper to draw all the rays on it instead of on the

HEIGHT MEASUREMENT 67

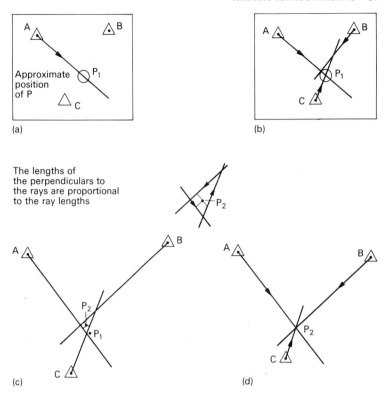

Fig. 3.13 Resection: (a) orient the table from the approximate position P by sighting to the furthest point A and draw the ray, clamp the table; (b) draw rays back from B and C; (c) the distances to A and B are nearly equal so the point lies midway between the rays and nearest to the line to C and the rays all go round P in the same direction; (d) turn the table until ray P_2B lies on B, then clamp and draw the rays back from A and C which will meet in a point.

plane table sheet itself. In this case we simply draw them outwards from a single point on the tracing paper by sighting on the fixed framework points without laying the alidade against their positions on the table. The tracing paper can then be moved about until the rays on it fit exactly over the plotted positions of the framework points on the table; the point from which they were drawn on the tracing paper can then be pricked through on to the plane table sheet. However, although simple, this is not the most precise technique and is difficult in a wind.

68 HEIGHT MEASUREMENT

It may therefore be preferable to work on the plane table itself and use the *triangle-of-error* method. In this case the three rays will normally fail to meet at a single point and will form a small triangle (Fig. 3.13(c)). The direction of each ray to its framework point should be marked with a small arrow, and the correct position of the occupied point should then be sought at a point where the rays all go round the point in the same direction and where it is nearest to the closest point and furthest from the furthest point in the same proportions as the lengths of the rays to them. Since the angles between the rays are correct, if they are to meet at the position occupied by the surveyor they must all be turned through the same small angle (which is the error in his orientation) and of course in the same direction, i.e. either clockwise or anticlockwise. If he is inside the triangle formed by the three points on the ground there is no problem because his map position will be inside the triangle of error; however, if he is outside the ground triangle then careful consideration will have to be given to which side he should be of the triangle on the plane table. If he has no triangle at all at the first attempt then he should be very careful indeed, because it may be that he is on the circle which passes through the three points and in this case the angles will be the same at every point on this and he cannot fix his position at all by resection alone. In fact, if the middle of the three framework points he is using is on the far side of the line joining the two outer ones he is in danger of obtaining only an approximate fix; they should either be nearly on a straight line or the middle point should be on the near side of the line joining the other two. In any case, of course, the largest angle between any two of the points should be at least a right angle or he will not obtain satisfactory cuts where the rays cross each other. Having selected his position relative to the triangle of error the process is now repeated, and the triangle of error should be very much smaller or should have disappeared altogether (Fig. 3.13(d)); if it is not then the selected point is on the wrong side of one of the rays. A second resection may be necessary to reduce the triangle of error to nothing, and if possible a completely independent check should then be made on a fourth point which has not so far been used in the resection. If this is done, and if the points are well distributed — preferably on all sides — then he can adopt the resected point as part of his framework and use it for further intersections and resections, but a station resected from only three points, or from four on one side only, should be used only for locating detail near it.

Heights can be observed with either a telescopic alidade, which includes a telescope carrying a vertical circle read against a bubble, or an Indian clinometer. The former is more accurate but together with its box weighs more than the ordinary alidade and clinometer together; moreover it reads directly in angles and so the tangents of these have to be looked up in a table or obtained by calculator. The Indian clinometer (Fig. 3.9(d)) has two folding vertical vanes; the vane at the eye end has a small viewing hole and the vane at the object end has a cursor which can be raised or lowered by a slow-motion screw and pinion working on a rack. This carries a short horizontal hair with which the distant object can be intersected or its summit touched; the tangent of the vertical angle observed can then be read directly off a scale at the side of the vane and a note made of whether it is positive (as elevation) or negative. The two vanes are fixed to a carriage which pivots at the front end of the base and can be made horizontal by turning a foot screw at the eye end until a spirit level in the middle is central; the object is then intersected and the reading taken. The tangent of the vertical angle is the ratio of the vertical to the horizontal distances between the occupied station and the object being observed, and so if the horizontal distance to this is scaled off the plane table and multiplied by this ratio the result is the height difference between the occupied station and the observed object, i.e. it is derived in the same way as by the Abney level (p. 55). Where an object not yet fixed is being observed the tangent is written against the end of the rays to it. When a second ray has been observed so that the object is fixed and the distances can be measured, the two height differences can be calculated. If the heights of the two occupied stations are known, then two values for the height of the newly fixed object are clearly obtained. Where a point of unknown height is occupied but has been fixed, as many values for its height can be obtained as there are points of known position and height visible from it.

When first used at the two initial occupied stations the clinometer should be checked by observing the vertical angle from both ends and seeing that there is no difference in the two tangents, although they will of course be of opposite sign. If the distance is less than 1 km care must be taken to observe to an object which is the same height above the ground as the plane table. Any large discrepancy must be removed by taking half of it out by adjusting the bubble. It will be clear that, since the clinometer is only about 20 cm long and there is no magnification or telescope, the intersecting accuracy will hardly be better than say, 0.25 mm or 1 part in 400 of the length of the observing base, so the angular accuracy is about 7' and we can expect height difference errors of the order of 2 m per kilometre. With this sort of accuracy corrections for the curvature of the earth and for the curvature of the light rays by refraction are not appreciable, since even at a range of 10 km the two together only amount to about 3'. It is only with surveys on a very small scale (of the order of 1/250 000 or less) and at ranges in excess of 10 km that this correction becomes appreciable.

The procedure is the same regardless of whether framework or detail points are being heighted: the point is fixed and then the height differences are obtained by measuring distance and vertical angles from the points with which it is intervisible. This is a very useful way of contouring; where detail points like bushes or rock outcrops can be intersected from known points then they can be heighted and contours can be drawn round them by eye from a distance. Where no such features exist the plane tabler will have to proceed from point to point resecting himself and drawing in the contours round each point as he heights it. In both cases the art is to choose significant points in the topography, for example those on stream beds, shoulders, hill tops, and changes of slope, so that the slopes between adjacent heighted points are fairly even. This type of contouring is inferior to and much slower than contouring from aerial photographs, and so it should only be used in fairly small areas where the expense of aerial photography and a proper photogrammetric survey are not justified or possible. It is also unsuitable for contouring flat areas where the useful contour interval is less than 1 m. The plane table can be used for traversing but it is not really suitable for this and it is better to use an instrument measuring angles and then record these, plotting the results afterwards.

Photography from ground stations

The correct use of a plane table not only requires graphical measurements of angles and some ground measurements of distance, but also the actual construction of the map in the field with all the mental effort this involves. In good working conditions with plenty of time plotting the map as one makes the observations has many advantages, but when conditions are harsh — because of cold, wind, rain, insects, or whatever — or when time is limited by frequent bad weather or other factors, then the mental effort and time required may not be available to do more than the absolute minimum. The observations and the plotting required to check them can then be limited with advantage to those required to establish the framework, leaving the details to be recorded by photography and plotted later in the comfort and leisure of a base camp or back at home. Instead of taking rounds of rays to intersect detail points the scientist can take rounds of photographs, but in order to do this properly some understanding of what is involved and adequate preparation are both essential. The simple geometry of the camera must therefore first be explained.

Although the aperture of the camera may be several millimetres in diameter, for our purposes it can be regarded as a pinhole through which rays of light pass in straight lines from the objects being photographed to create images (which will be upside down and reversed) on the negative which is held flat against the *format* of the camera. This should be in a vertical plane, i.e. the camera must be levelled, and the lens focus should be set to infinity. The scientist plotting from an enlargement made from this negative will be in the position of a plane tabler working from the station where the photograph was taken and looking through a window which subtends the same angle at him as the format did at the lens of the camera. Instead of sighting on the features he wants to intersect with his alidade, he measures their distances from the centre of the photograph and then draws rays through them on his plane table. However, we must be more precise than this. The scientist must know the following two geometrical characteristics of the camera and the resulting enlargement (Fig. 3.14):

(i) the position of the *principal point*, which is where the axis of the lens meets the format, or in other words the foot of the perpendicular from the lens centre (strictly speaking the rear *node* of the lens – see Fig. 7.1(b)) to the plane of the format;
(ii) the *focal length* when the lens is set to infinity, i.e. the length of this perpendicular and of course its equivalent length for the enlargements.

The principal point is normally near the centre of the photograph, and is likely to be very nearly in line with the midpoints of its shorter sides – what we may call for convenience its vertical sides. However, because of problems in accommodating winding mechanisms, etc., it may not be in line with the mid points of the longer sides, which are normally horizontal, and this must be checked by *calibration* unless the manufacturer can give a categorical assurance of its position on the format. The focal

72 HEIGHT MEASUREMENT

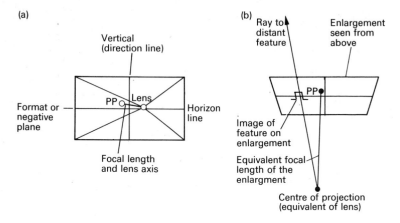

Fig. 3.14 The geometry of (a) the camera and (b) an enlargement.

length is normally recorded on a lens, and with 35 mm film cameras is normally 45 mm; however, a wider-angle lens with a focal length of 30 mm or so will have advantages and should be used if possible. To determine the equivalent focal length of the enlargements their dimensions must be compared with those of the negative. To calibrate the camera, which will check on both these characteristics, the scientist should find a suitable station with a good view of some reasonably horizontal line, such as the far shore of a lake or river, or a plain, with about a dozen well-marked sharp features on it which can be intersected on the plane table and also recognized afterwards on the enlargement (Fig. 3.15(a)). He photographs these with the camera set as level as possible on the plane table or on its own tripod, and then uses the alidade to trace on a piece of tracing paper stuck with scotch tape onto the plane table a ray to each of them from a single point (see Fig. 3.15(b)). When the photograph has been developed and enlarged he then draws lines on the enlargement joining the midpoints of its four sides; these will represent the horizon and the vertical at right angles to it, and their intersection will define the centre of the enlargement and should be near the position of the principal point. If the camera was level, all the features chosen should lie along or very near the horizon line (Fig. 3.15(a)) and this is the first check that the format was vertical when the photograph was taken. Distances on the enlargement from the vertical line are then plotted on

HEIGHT MEASUREMENT 73

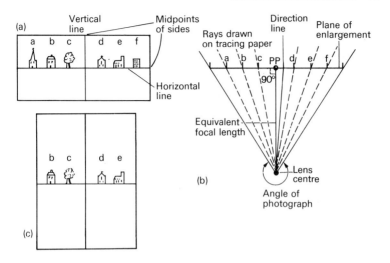

Fig. 3.15 Calibrating the camera: (a) enlargement; (b) resection of the lens centre and the determination of the principal point (PP) and the equivalent focal length for horizontal angles; (c) determination of the principal point (PP) for vertical angles.

a straight line drawn on a separate sheet of paper, and the rays drawn on the tracing paper are placed over this so that each of them passes through its own feature. The scientist will realize that he is in effect resecting the position of the lens centre relative to the plane of the format from a number of points along a straight line representing it, and it will be clear why a wider-angle lens will give a better result with a better angle of cut. Having pricked through the position of the lens centre from the tracing onto the paper, the perpendicular from it is drawn to the line. The foot of this perpendicular now represents the position of the principal point on this line and can be transferred from it back to the horizontal line on the enlargement, while the length of the perpendicular is the equivalent focal length of the enlargement.

Strictly speaking, and in order to be able to obtain vertical angles as well as horizontal ones, the calibration should be repeated with the camera turned through 90° about the lens axis, although in this case levelling the camera may be more difficult and the width of the angle with a rectangular format, and consequently the number of features, will be less and so the result will be less accurate (Fig. 3.15(c)). Once both these calibrations have been carried out — and this is best

74 HEIGHT MEASUREMENT

done before starting in the field — the exact position of the PP can be marked on the calibration enlargement and transferred from this to all other enlargements. At each survey station in the field the scientist then levels the camera and points it at a feature which he can intersect on the plane table, takes the photograph, and draws a ray on the plane table (which of course has been correctly oriented) to the feature, labelling it also with the number of the photograph. Successive photographs are taken and their directions recorded in the same way, giving a good overlap of 5° or so at their edges. The preliminary calibration will have shown him how wide an angle the camera covers. Ideally, the film should at least be developed, and enlarged also if possible, in camp and before field work is finished; thus any bad exposures can be repeated. This is of course also a good procedure in bad weather areas because time confined to base can then be spent in plotting the results, and any areas not properly covered can be seen in time to cover them before the field work stops.

Plotting is very simple (Fig. 3.16). The directional rays of the photographs are already recorded on the plane table; the equivalent focal

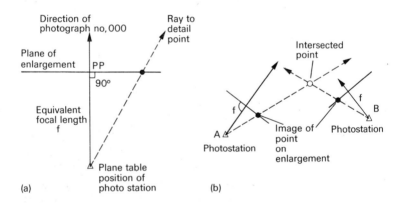

Fig. 3.16 Intersecting detail points from camera stations.

length of the enlargements is set out along each ray and a line is drawn at right angles at this distance from the station point to each ray. Distances along (or parallel to) the horizon line on the photograph from the PP to any detail points to be fixed are transferred (by measurement or better directly with dividers) from the enlargement to the line and rays then drawn through these from the station position;

clearly they will be equivalent to rays drawn directly on the plane table in the field. Similar rays are drawn from other stations in the same way (Fig. 3.16(b)) and the intersections will mark the positions of the detail points, with useful checks provided by three or more rays to the same point. Apart from the advantage of being able to do this work in comfort and at leisure instead of on some cold, windy, or insect-infested hill top there is another tremendous advantage over field intersection, particularly in small-scale work when a day or more may be required to travel between successive stations. This is that a much greater number of detail points can be intersected because it is much easier to recognize and identify them from the two or three stations when all the views are available simultaneously than when trying to remember at the second or third station in the field which points had been intersected from the first. In addition to obvious features like hill tops or buildings, when viewing the photographs one can pick out and recognize boulders, trees, or even vegetation, snow patches, crevasses, or changes of shade or rock on the sides of hills which could not be recognized from two different survey stations in the field even if the most elaborate panoramas were drawn from them. If both calibrations have been done as described above and if the camera tripod enables it to be accurately levelled, then vertical angles can be measured graphically in exactly the same way, and so heights, and hence contours, can be determined in this way just as in normal plane tabling with a clinometer.

To sum up, the scientist going to an area where the weather is likely to be bad or working conditions harsh should consider seriously *before he goes* whether his plane table work could be usefully augmented by photography, and if so he should equip himself properly for this. The camera should be capable of being levelled accurately, either on the plane table or on its own tripod. It should have a normal or wide-angle (but not a 'fisheye') lens. It should be calibrated before leaving. The directions of all photographs must be recorded in ink on the plane table and annotated with their numbers, and a separate list should also be made. If possible one photograph at each station should include at least two features which have been intersected, and their vertical angles recorded, on the plane table; these will act as a check on the levelling and orientation of the camera at that station. The enlargements should be made as soon as possible, preferably in the field before work ceases. If time allows plotting should also be done in camp or at base. Enlargement should all be done with one setting of the enlarger; although differences can of course be taken up it will be much easier if they do not have to be.

The scientist who is going to attempt the more advanced type of surveys described in the next few chapters, where a theodolite is used, will realize that this kind of terrestrial photography can also be done in conjunction with detail survey by theodolite, and is more accurate if more trouble is taken and the camera is calibrated more accurately, especially if the format is fitted with *collimation marks* to define the principal point exactly and the camera is set up precisely over the theodolite or on its tripod so that it can be accurately levelled. Such refinements, and also the use of specially made *photo-theodolites*, are really beyond the scope of this book, but references are given to which the interested reader may refer. In addition, if he masters the theory and practice described in the part of the book dealing with the use of stereoscopic pairs of vertical air photographs he will realize that the same principles can also be applied to the use of stereo pairs of horizontal photographs taken from ground stations particularly in rugged country; this also has a particular application for recording and measuring the shapes of buildings and other man-made structures such as components of ships being constructed in separate yards, oil rigs, large bridges, and so forth. Another use is when time is limited, e.g. in recording the details of a road accident or surveying inside a cave.

4. OTHER CONSIDERATIONS

Errors and specifications

On pp. 34 and 35 we considered briefly the types of errors which may affect a traverse. The time has now come to look at these in more detail in relation to the proposed uses of a survey or map and how these requirements can be expressed in an accuracy specification. Measurement errors fall into three categories: accidental, systematic, and gross. Surveying errors can be divided into two categories: those in angles or bearings, and those in distances.

Accidental errors

Accidental errors approach the precision of reading a measurement and are by definition equally likely to be positive or negative. Examples in angular measurement are errors in estimating the tenths of the smallest reading division of a circle or micrometer graduation, or in identifying the coincidence of graduations on a vernier. However, other errors may be present in the actual engraving of individual divisions since nothing is perfect. In more precise work the refraction of the line of sight as it passes through the atmosphere may also be appreciable, and it is for this reason that very precise observation of angles is spread over one or more days in order to eliminate bias due to, for example, morning or evening conditions. In distance measurement similar errors occur both in reading and in the graduation of tape scales, while in more precise work further errors arise in graduating and reading the thermometer and the spring balance which record the temperature and the tension respectively of the tape.

The effects of accidental errors are minimized by repeating the observations as far as possible in different but equivalent conditions. In measuring a single length the tape should be slackened and pulled tight again for each repeated measurement, and in precise work the distance to be measured is less than the full tape length so that each of the successive readings at the ends is made on different divisions of the scale. To measure a distance of several tape lengths the pegs or arrows between which each tape length is measured should be differently placed for each measurement of the whole. As mentioned

78 OTHER CONSIDERATIONS

earlier (p. 34), when the final answer is the sum of several quantities true accidental errors accumulate as the square root of the number of these quantities. In a series of measurements of a single quantity (such as an angle or a tape length) the final answer is normally the average or *mean* of these which is obtained by dividing their sum by the number of measurements, and the difference between the largest and smallest measurement is the *range* of the observations. In a single 30 m tape length, for instance, the range would be a centimetre or two under good conditions and in a compass bearing read to $0.5°$ the range of several readings will be a degree or two. If several readings are taken, say half a dozen or more, it sometimes happens that a gratifyingly small range can be achieved if one reading is omitted, while if it is included the range may be doubled, giving an apparently less accurate mean. In trying to achieve extra precision by repeated measurement of the same quantity it is a common and dangerous fault to reject any 'rogue' readings which differ in this way from the mean by considerably more than the average. Accidental errors, as already explained, usually include more than one source — of manufacture and environment as well as observation — and while these often balance out to produce a relatively small individual variation from the mean, they may sometimes all go one way in an individual measurement producing an apparently 'rogue' result. Rejecting these in calculating the mean may well bias this and give the surveyor a rude shock when he applies an independent check, for example by adding up the three angles of a triangle in which this kind of selection has been applied in deriving the mean values of each angle. However, it is always possible that an individual observation really is grossly incorrect and should be rejected. It is a good rule of thumb to calculate the mean using all the observations, and then to compute the average difference of individual observations from this. Any individual observation which differs from the mean by more than three times this average difference can then be rejected; any with smaller differences should be retained in calculating the final mean.

Systematic errors

A distinguished surveyor and scientist once said:

> Every step which removes a part of an error of observation from the accidental category to the systematic is a real step forward.†

† G. P. Lennox-Conyngham, Foreword to *The arc of the thirtieth meridian between the Egyptian frontier and latitude 13° 45'*, by R. C. Wakefield and D. F. Munsey, Sudan Government, Khartoum, 1950.

This should be the aim of every scientist. In surveying, the way in which systematic errors arise can be seen more easily than in some sciences where the measurement is less direct. Systematic errors are those in which the observer, the instrument, or the environment have produced a bias. If so, however often the observations are repeated the end result is still likely to differ from the true value by an appreciable amount which is greater than the precision or reproducibility of the observations. In distance measurement errors arise from using a tape which is the wrong length (either in manufacture or more commonly due to a badly repaired break) or which is being pulled with a consistently wrong tension; in more precise measurement errors can arise if the temperature is being consistently wrongly observed. Anything which affects the straightness of the tape must have a systematic effect since it will always give rise to a value which is greater than that of the straight-line distance; examples are poor alignment on the ground, sag in unsupported sections, and unrecorded slopes. In more precise work a faulty spring balance or thermometer, or a constant difference between the temperature of the thermometer and that of the tape, can produce systematic errors. Systematic errors are less likely in simple angle measurement, but in precise work using a theodolite the observing procedures must take full advantage of the design of the instrument to balance out any mechanical faults such as eccentricity of the circle, lack of perpendicularity of the axis of the telescope to the two axes of rotation, and so on. Lines of sight passing near a heated object such as a rock or building will be bent and cause a systematic error. In levelling, the use of sights of unequal length, which can save time going up or down hill, can have a systematic effect particularly if the level is out of adjustment, and the sinking of the level or the staff into soft ground between successive sights will also have a systematic effect. The result will be a disappointingly large closing error when a circuit of levels is closed.

Gross errors or blunders

Gross errors are most likely to occur in repetitive and boring operations like long traverses, base measurement, or spirit levelling. Whole tape lengths can be miscounted, distances misread by metres or ten of metres, or where surveyors are reading at the two ends simultaneously and only one is recording the recorder may mishear a reading.†

† I have been told that in a very precise survey for a motorway bridge in Britain the repeated mishearing of 25 for 29 by one surveyor caused an error of 4 m in a section of the measured width of the river which was only discovered by an independent direct measurement using electronic techniques.

Angles can be misread by a degree or by tens of minutes, or on the wrong side of a graduation, or of course to the wrong object. Only unremitting care and constant checks can reduce these errors to the minimum. Examples in tape measurement are moving the tape lengthways so that quite different readings are taken at the ends for each measurement, using two different tapes (or both sides of the same one) graduated in different units such as feet and metres; reading the stadia hairs in a level telescope as well as the centre hair; and using the moveable horizontal circle of a theodolite to obtain completely different readings for the individual points. These are all checks to be taken in the field at the time of observation and to be worked out on the spot before moving the instrument, but further and completely independent checks should also be applied by what are called redundant measurements — the three angles of a triangle when two will define its shape, the diagonals of a taped quadrilateral as well as the sides, closing traverse and levelling circuits, measuring the distances between individual points fixed by bearing and distance, and so on.

Relative and absolute errors

So far we have been concerned only with errors in the actual observations, but the end result of all surveying is to produce either a map or a structure set out on the ground from a map in which individual features or points are located by two (or three if heights are included) dimensions or co-ordinates. This brings in a further division of survey errors into *relative* and *absolute* errors. In general it is the relative errors between neighbouring features which matter to the map user since these are the ones he can see for himself if they are large enough, while only more elaborate survey techniques will disclose the absolute errors which occur when a whole section of the map is misplaced relative to the framework or the grid on which it is based. Errors can be expressed in two ways, as a proportion of a length or as an actual displacement on the ground or on the map, and it is important to see how these interact with each other especially when a specification for a survey is being considered. As an example of this it is clear that a gross error of 1 m in the length of a taped traverse would be an unacceptable relative error between the two ends of the leg (say 100 m long) in which it occurred, yet might not be appreciable over a circuit of 1 km or more of which that leg was only a small section. In contrast, if the accidental closing error of the traverse circuit was about 1 m and this was adjusted through the individual legs, the individual errors in these would be only a few centimetres which would be acceptable relative errors between their ends. However, the absolute sideways error of position of the middle of the traverse even after adjustment might well be about a metre, and all the detail fixed from it would have a similar absolute error in position. This would only matter if another long traverse, also with an error of a metre in the middle, possibly in the opposite direction, passed near the first one without being connected to it. Then the two absolute errors would cause a large relative error over a small distance, which underlines the importance, already mentioned on p. 40 of not allowing this to happen, however hard it may be to put the connection through in cases where no path or track joins the two routes along which the two traverses have been run. We can see from this how important it is in specifying the accuracy of a survey framework or a map, or a layout set out on the ground, to state clearly not only the maximum relative error allowable between neighbouring points or features but also the proportional errors allowed across the whole survey.

Specifications

Despite the above commonsense statement, it is surprisingly common to read a specification saying that a large-scale plan must be accurate to (say) 0.5 mm, or to a few centimetres on the ground, without saying over what distance this must apply. If the area being mapped is large, say 10 km in extent in any direction, then an allowable error of only 10 cm over the whole area would mean that the framework had to be accurate to about 1 part in 100 000, which would require very sophisticated and expensive survey techniques, while what the author of the specification probably meant was that the distances between neighbouring features only a few metres apart must be accurate to 10 cm, requiring a proportional accuracy of only one part in a few hundred which is achievable with simple taping techniques. For most readers of this book the sort of maps they make and use will be adequate for their purposes provided that the relative errors between two features a few centimetres apart on the map do not exceed a millimetre, and that the absolute errors of a single map sheet do not exceed a centimetre or so. Thus in moving over the map and fixing themselves or other features on the ground by simple means such as a tape and compass they will find that checking on several different features shown on the map will always produce very much the same positions to within a millimetre or so.

Great accuracy is always expensive since it requires more expensive and sophisticated equipment and staff. A scientist laying down a specification for a survey to be carried out by himself or his colleagures, or by professional surveyors, should look very hard at the uses to which the results are going to be put, not only in the immediate but also in the more distant future. In particular he should consider the precision with which the features to be mapped can actually be defined and the accuracy to which other quantities related to these can be measured. Clearly buildings whether old or new can be defined to within a few centimetres and the map accuracy may have to reflect this, especially if they are permanent; however, soil-type boundaries can seldom be defined more precisely than to a metre or so on the ground, and a map on a very large scale showing them to the nearest 10 cm on the ground is a waste of effort. The scientist should also bear in mind that while he may want to label the individual plots (or houses) and needs room for this on the map, he may not require their boundaries to be accurate at the large scale that this would require when the plots are small. The answer in this case is to survey at a smaller scale and then enlarge the result photographically, accepting that the boundaries will then be rather coarser than they would be on a well-drawn map.

A good example of where other quantities are involved, as well as the areas or slopes to be taken from a map, is given in the four types of map normally required by engineers designing an irrigation scheme, including the reservoir to supply it with water. They require maps of the catchment basin of the river, of the proposed reservoir area, of the actual dam site and main canal headworks, and of the area to be irrigated. Each requires a quite different specification. The map for the catchment is for calculating probable river discharges in normal and extreme conditions; this can be on quite a small scale and coarsely contoured since the other quantities involved, i.e. the average rain fall, the evaporation, the seepage, and the run-off, are unlikely to be known or estimated to nearer than about 10 per cent of their true value so there is little point in knowing the areas or slopes to great accuracy. The reservoir map must be on a larger scale and fairly closely contoured since the engineer requires to know the relation between the height (and cost) of the proposed dam and the volume of water it will hold back. However, this is not all that precise because wind can affect the surface slope (especially in long reservoirs) and from the moment the dam is closed the reservoir volume will start to decrease by silting at a rate which can only be estimated approximately, so once again the uncertainty of the other quantities involved makes too great a scale and accuracy, and too close a contour interval, unjustified.

The greatest precision, and the largest scale, are required for the site of the main engineering works, i.e. the dam and head works of the main canal. In this case the engineer will want to calculate the volumes of earth and rock to be removed, and the concrete structure to be designed and installed; he may well require maps on a scale of 1/1000 or even larger, with contours at 1 m or 0.5 m intervals, and a very accurate and permanently marked framework from which to set out the structures and for monitoring the distortions of the dam as it comes under load. Downstream, for the irrigation scheme, a smaller-scale map will be acceptable, but clearly the heights must be very precise since the area is flat and an effective canal network has to be designed and set out; obviously good permanent benchmarks are also required here for this purpose. For very large dams the sort of scales and contour intervals required for these four maps would be about 1/50 000 or 1/100 000 for the catchment with contour intervals of 10 m, 1/20 000 for the reservoir with contour intervals of 5 m, 1/1000 for the dam site etc. with contour intervals of 0.5 or 1 m, 1/10 000 or 1/20 000 for the irrigated area also with contour intervals of 0.5 or 1 m.

One other point affecting accuracy should be mentioned: this is the size not so much of the area on the ground but on the map, and particularly whether it will cover more than one map sheet and whether more than one scientist or surveyor is likely to be working on its compilation on the ground. In this case, as is described in more detail in the next chapter on advanced surveying, the accuracy of the framework would, simply for practical surveying reasons, have to be greater than that required in the maps themselves, otherwise there will be difficulty in reconciling both the work on adjoining map sheets and the work in adjoining areas surveyed in detail by different people. Errors of absolute position of a centimetre or so at map scale will in many cases cause little trouble to the scientist using a single map sheet, but unless the features shown are very indeterminate, they would cause difficulty when the same feature fails to join by this sort of amount across a sheet or working area edge and the framework over the whole area must be accurate enough to prevent this happening.

The limitations of simple surveys

If the reader has digested the preceding two chapters and practised the techniques described in them — preferably with the help of someone (not necessarily a surveyor) with previous experience — then he or she should now be able to make a satisfactory map of any area of relatively small extent with any of the characteristics (or a combination of them) described in the introduction. For many purposes the internal accuracy of a map need be no higher than that of visual estimates of distances and angles, because a person using it without any measuring instruments will then be unable to detect any errors. This means that there is a local and relative accuracy of about 1 part in 50, and this low accuracy is adequate in many kinds of scientific field work because the features being mapped cannot be defined accurately. For example, soil and vegetation boundaries tend to be indefinite and to have a width of several metres on the ground and to be located in different places by different individuals. Nevertheless, in order to preserve even this local accuracy in the detail survey we have seen that the framework over the whole area has to be surveyed more precisely, or there will be difficulty in fitting the different parts of the detail survey together. Moreover, for many purposes, especially in engineering, very much greater accuracies are required even than the 1 in 1000 obtainable by plane tabling or careful ground taping; structures up to 100 m or more in extent have to be designed and built and fitted into place with accuracies of millimetres which requires a proportional accuracy of 1 in 100 000. In general engineers concerned with such structures have the mathematical ability, and in some cases the survey training, to be able to do the survey and setting-out work themselves, and this book is not intended for them.

For those dealing with less precisely definable features the preceding sections may seem to contain all they need to know. However, this is not really so, and the first step in appreciating this is to look more closely at the accuracies which can be achieved by simple techniques and at the limitations on them even when it is assumed that the conditions are ideal and that all possible care has been taken. On flat ground a chain or tape survey can be as accurate as the accuracy with which the tape length itself can be measured, since the actual measurement can be to the nearest millimetre (or even less if a magnifying glass and a very fine mark are used). The real errors come from incorrect tension of the tape, or to a lesser extent from changes caused by varying temperature or humidity, from incorrect alignment either vertically or horizontally, and finally from the tape not being horizontal or from its slope not being correctly measured. Accuracies of the order of one part in several thousand can be achieved quite easily by using a steel tape or a fibreglass tape correctly tensioned with a spring balance, and by careful alignment and levelling on clear flat ground, and of all the techniques we have considered for planimetric surveys this can be the most accurate. Its accuracy can easily exceed that of plotting and drawing a single map sheet, which as we have seen can scarcely be better than 0.5 mm in 50 cm or about 1 part in 1000. This means that to obtain the full value from a really accurate tape survey we have to express the results as co-ordinates rather than plot them directly, since even at a very large scale such as 1/10, with the map or plot spread over several metres, the dimensional stability even of plastic map material will not be good enough to ensure accuracies to much better than a few millimtres over the whole plot. Thus taping without angular measurement can be a very accurate form of survey, but obviously only over a limited area which must be flat and free of obstructions.

As we have seen, if care is used and if the ends of the base line and the first framework points to be established are near the edges of the map sheet, a plane table survey can be accurate everywhere to about 0.5 mm, giving a proportional accuracy of 1 part in 1000. Since it is the map sheet itself which is the measuring instrument there is no way that we can improve on this however hard we try. If therefore a plane table survey is to extend over several map sheets and is everywhere to retain plottable accuracy relative to a plotted grid, it must be controlled by a framework of higher accuracy for which more precise distance and angle measurements are required using more sophisticated instruments and techniques. The plane table is preeminently a topographer's instrument for use in open hill country, but it is also one which the field scientist will find very useful for mapping features not shown on existing maps or those which cannot be seen from a distance and which therefore have to be followed on the ground with successive points along them being fixed by resection. It is not suitable for very large scale surveys (larger than about 1/1000), because the size of the table itself then becomes appreciable and it has to be set exactly over each occupied station so that not the centre of the table but the position of the station on it when it is correctly oriented lies over the position of it on the ground. Moreover, unless

86 OTHER CONSIDERATIONS

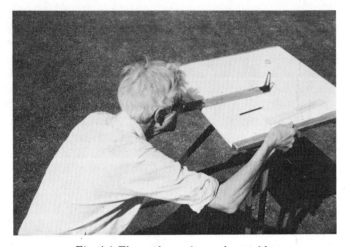

Fig. 4.1 The author using a plane table.

the features being fixed as detail points are very precise, an assistant with a ranging rod has to be employed to define them with a precision comparable with that with which they can be plotted on the table. This turns a simple technique into one of some complexity and removes most of its advantages. The plane table is also unsuitable for use in forested country and for running traverses, especially if the legs are short. Angular errors soon accumulate and cause a swing of the traverse, and it is better to orient the table by compass at each set-up. However, if we are going to do this we might as well use the compass directly for the traverse angles.

The compass survey is the least accurate of all, both because of the inherent lack of precision of the compass itself and because of the possibility of changes in the magnetic declination both over an area and during the day. Some improvement can be obtained by using a large stand compass set up on a tripod and repeating each reading several times, but even then the accuracy is unlikely to be better than about 0.25° or 1 part in 200. Over a whole map sheet this could produce errors of 3 or 4 mm, and although for some purposes this may be perfectly adequate the result is not an accurate map. If the area to be mapped extends over several map sheets and these are surveyed at different times and by different individuals, trouble will be experienced in fitting them at the common edges. Compared with this limited accuracy the compass has several advantages, particularly in close country in an area of magnetic stability. Its main advantage then is that it always gives a constant direction, and so the accumulation of bearing errors along a traverse is minimal. The other great advantage of the compass is its lightness; even a large stand compass with its light tripod and a tape, or a pedometer or a wheel, can all be carried easily by the scientist himself all day without undue fatigue. Thus it is particularly suitable for detail survey by traverse or locating lines of levels if controlled by a good framework or for the preliminary reconnaissance of an area for planning a more accurate and detailed survey.

In height measurement the spirit level is by far the most accurate instrument, and in careful hands even quite a small instrument can give very good results of the order of a centimetre times the square root of the length of the level line in kilometres. Without using a theodolite the accuracies of heights determined by distance and vertical angle are of a very much lower order, but the right answer may be to establish a higher-order framework of benchmarks by levelling along the flatter roads and paths round and across the area and then breaking this down and measuring detail heights for contouring with a plane table and clinometer or with a compass and Abney level. As has been stated earlier, the overriding consideration must be whether or not the uses to which the map will be put include the design studies of open water levels for drainage or irrigation.

While relatively few non-surveying scientists will have time to spend on more accurate and extensive surveys than those so far described, this is no reason for not including them in this book. The following chapters describe how greater accuracies can be achieved in both distance and angular measurements by any competent scientist who has an interest in doing so and is prepared to learn how to use a theodolite and get the best out of a steel tape. These chapters will also give some idea of the even more accurate and sophisticated techniques used by professional land surveyors so that the scientist will have some idea of what they are talking about if he is working alongside them on an extensive project, and also, if he is senior enough to be involved in negotiating a survey and mapping contract, to know enough to suspect that there may be weaknesses in some of the proposed methods and techniques or those being used on his project. Contractors are in business to make money, and this normally involves compromises of one kind or another, while local departments in developing countries may wish for prestige reasons to undertake work beyond their real capabilities; those who will be using the results of the contract should have at least some idea of which economies are acceptable and which are not. The main difference between the surveys described so far and those to be considered in the next chapters are that they have involved only some simple arithmetic and that the data have been plotted directly as distances and bearings on the map. In more precise work the distances and angles have to be converted by trigonometry into rectangular co-ordinates so that the points can be plotted on a series of gridded map sheets. Not only is this a more accurate method of plotting each point because the lines defining it are at right angles, but it retains its accuracy relative to the grid over the whole extent of the map area. Until a few years ago these trigonometrical calculations were rather tedious and required the use of logarithms or hand calculators, and mistakes were easily made in the arithmetic, and in looking up the logarithms or the values of trigonometric functions in books of tables. Nowadays a small electronic calculator costing a few pounds or dollars will do all this work for the surveyor at the touch of one or two buttons and will produce the correct result provided that it is given the correct data. We shall also see in the chapters on air survey how quite extensive and accurate maps can be based on a relatively open framework of points fixed on the ground and identified on a few air photographs, and that the field work required to do this may be so limited that it is worthwhile overseas for the scientist to do it himself rather than to pay for flying out a surveyor for possibly only a week or two's work. If the scientist is looking for one simple rule of thumb to help him decide whether he can tackle a survey himself or should call in a professional land surveyor he should ask himself: Is the area so extensive or is the required accuracy so high that the curvature of the earth has to be taken into account in designing the grid (as for example in all national grids) and in computing co-ordinates on it? If this is the case then he had better leave it to the professionals, but if not, and he can work on a flat earth (or on a small enough part of a national grid to use average values for scale and convergence), then there is no real reason why he should not be able to do it himself.

Simple hydrographic surveys

The next two sections outline solutions for specific problems which may arise in the study of surface, underground, or submarine features. Their solutions include a variety of techniques and instruments which, with one important exception, have been described in detail already. The exception is the *sextant*, which was specifically developed for use in a vehicle in motion as compared with all the other instruments which have been developed primarily for use by an observer standing firmly on solid ground. The problems covered are

(i) surveying surfaces under water from the water surface above them (bathymetric or hydrographic surveys);
(ii) surveying underwater features or structures from the bottom while under water oneself (submarine surveys);
(iii) surveying caves and tunnels from the inside;
(iv) measuring the heights of individual objects.

Hydrographic surveys

It is a basic fact that all surveys made from a boat are slower, more difficult, and less accurate than those made on land, except in a marshy area or in dense forest, mainly because the actual transportation and working conditions of the surveyor are more complicated. Any reasonably numerate person with two legs and two arms (and even one if sufficiently determined), or in more extensive work with the ability to drive a Land Rover or similar vehicle, can learn to work as a field surveyor on land; but almost all hydrographic work requires the ability to manage a boat (or to employ someone else to do it), and to become an efficient sailor is not learnt in a day. Moreover, the movement and acceleration of a boat, and the uncertainty of its position at any moment, make accurate survey work from it much more difficult than from a fixed land station. As against this, a lower degree of accuracy is normally acceptable in hydrographic work because the user himself is less sure of his position. Thus in planning the hydrographic survey of the bottom of a body of water with a varying surface level it will pay to do as much of the work as possible by ground methods when the water level is at its lowest: in lakes following a dry spell, and in tidal areas at low spring tide. Where a reservoir is planned it will be far easier and cheaper to survey the area to be submerged before it is filled than afterwards. In many parts of Africa quite large rivers dry out completely for part of the year or are confined to a narrow channel, and this is the best time to measure their cross sections or contour their flood plains.

Having planned the survey to cover as much as possible on dry ground, the next step is to provide an accurate map of the coast line by normal ground survey methods including a series of features which are accurately fixed in position and will be clearly visible from a boat on the surface of the water. Fixed points out in the middle will only be required in extensive areas normally beyond the capability of a field scientist who is not a surveyor. If some already exist, such as islands, rocks, beacons or even mooring or navigation buoys, then they should be fixed in position from the land by triangulation or intersection. Mooring buoys will of course change position with varying tidal currents, particularly just before and after low tide when their cables are more nearly horizontal; but for many hydrographic purposes the amount of movement may not be significant since the accuracy required of hydrographic surveys is lower than for those on dry land. All these features form the framework of points which is as essential in hydrographic work as in all other types of survey, but because less accuracy is required specially beaconed points are often unnecessary and features such as conspicuous buildings or trees will be adequate.

Normally the main purpose of hydrographic surveys is to provide contours of the bottom, though observations of its nature and of any special features such as wrecks or archaeological remains may also be required. The work divides into first the fixing of the scientists's position and then taking soundings or other observations. For fixing his position the procedure is similar to that used in providing spot heights when levelling a flat area, i.e. the running of relatively straight parallel lines between fixed points with soundings at regular intervals along them. If the lines are long, a number of checks on position must be made at longer intervals. In inland waters without tidal streams and in relatively calm conditions this procedure is not difficult. It requires a separate competent person to row or steer the boat on a straight course and at a regular speed, so that the soundings can be made at known intervals either by timing the intervals or by counting the number of strokes of the oars between soundings. If this is done properly then the position of any point along the line can be interpolated on the map between the two ends of the line as in adjusting the individual stations of a traverse or level line.

OTHER CONSIDERATIONS 91

For any but the smallest areas some sort of separate position checks are required, both because of the likelihood of a bend in the middle of the line and because of possible unevenness in the spacing of the soundings along it. In calm conditions this can be done by using an

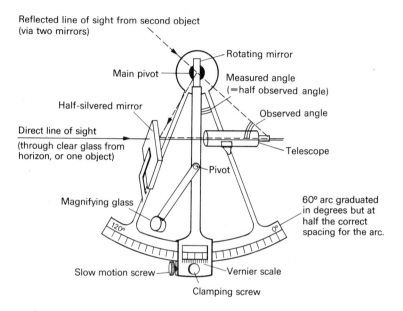

Fig. 4.2 The sextant.

ordinary prismatic compass (or a special hydrographic compass) and taking bearings to three or more of the framework features on the shore. If possible these should be all round the point being resected, but if not then the two bearings furthest apart should subtend an angle of at least 90°. Elevated points such as hill tops can be used with a compass provided that they can be seen in the sights. However, greater accuracy requires the hydrographer's equivalent of the land surveyor's theodolite — the *sextant*. Because his observing station is moving all the time it is clearly impossible to use any instrument like a theodolite which measures an angle by first pointing at one object and then at another and rotates a reading device above a supposedly fixed and immovable graduated circle; the sextant solves

92 OTHER CONSIDERATIONS

this problem by using a system of mirrors to observe simultaneously the two objects which define the arms of the angle required. Figure 4.2 shows how the sextant works by bringing together in the telescope eyepiece the direct ray from one object and other ray reflected in two mirrors of which one is attached to a rotating arm carrying an observing device which traverses a graduated arc. Because the rotation of the second ray is twice that of the mirror, the graduations of this arc are at half the normal interval, and although the arc itself subtends an angle of only 60° (hence the term sextant) its graduations will cover twice this or 120°. Reading is by either a vernier or by a simple micrometer, and normally is to not better than 10".

For accurate work it is just as important as on land that the angle measured should be horizontal, but in a moving boat it is impossible to level the sextant with a bubble as one does for a theodolite. This is overcome by using points on the shore at water level or very near it; elevated points such as hill tops must not be used for horizontal angles with a sextant or a false angle will result. For each fix at least

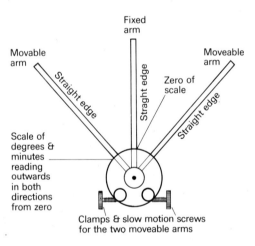

Fig. 4.3 Station pointer (for drawing two angles observed at one point).

two angles are required between three points on the shore. Either the boat must be stopped for their measurement, or two observers with two sextants are required which is the normal practice in professional hydrographic survey work. However, one sextant will be as much as the amateur is likely to be able to borrow, and he should if possible obtain a special sounding sextant with a wide angle telescope rather than one designed for astronomical fixes since this is more difficult to use quickly and the extra accuracy it can provide is not required. Horizontal angles to the nearest minute are more than adequate since they cannot be plotted on the map to better than this. Plotting can be carried out either by setting out the angles with a large protractor on a piece of tracing paper and fitting this over the three points observed, or by using a *station pointer*. This consists of three arms pivoting about a point and a graduated circle by which the straight edges of the outer two can be set at the two observed angles on either side of the middle one (see Fig. 4.3).

If an object of known height, such as a lighthouse or flagpole, is sited on the shore so that both its top and bottom can be seen, then a distance can be obtained by measuring the vertical angle which it subtends at the point to be fixed; however, this technique is limited in use and not very accurate. For example, if the sextant can only be read effectively to a minute of arc and the object is 30 m high, at a distance of 1 km it will subtend an angle of only 30/1000 or 1/33 which is about $1.7°$ or $60 \times 1.7 = 102'$. This can only be measured to $1'$ or 1 part in 100, so the accuracy of the distance at best is about 10 m (or 1 mm on a 1/10 000 map). The error increases as the square of the distance so that at 2 km it will be at least 40 m. In fact it will be more because there may well be a change in refraction between the two rays to the top and bottom of the object, unless the air and the water are at nearly the same temperature. This method should therefore be used with caution and only when a normal resection is impossible. Clearly the exact position of the bottom of the object must be correctly defined; if it is water level then any variations in this such as the tidal range must be allowed for. The distance is obtained by dividing the height of the object by the tangent of the observed angle, either using tables or nowadays by using a small hand calculator (see Fig. 4.4).

Once the position of the observer on the water surface is fixed by these means the investigation of the nature of the bottom or of a wreck or other object of interest is not a survey problem; however, measuring the height of the bottom is the other half of hydrographic survey. The fact that the observer is working from a level surface above it is clearly an advantage, and all he has to do is to measure the depth below this using either an echo sounder or a simple lead line. However, in tidal waters the level surface is rising or falling, and

94 OTHER CONSIDERATIONS

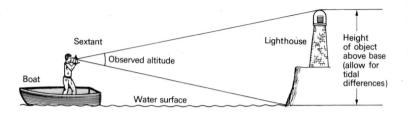

Fig. 4.4 Observing a subtended vertical angle to derive distance.

all observations of depth must be timed and corrected for the state of the tide by comparison with local tide tables (if these exist) or by timed observations at regular intervals of a *tide gauge* set up on shore. Similar precautions may be required for observations made in a lake or reservoir if they are extended over a long period. The gauge can be a simple graduated pole set so that its bottom is never exposed and its top is never submerged, or some sort of automatic water level gauge using a float and recording drum. Even when local tide tables are used some check is required because they may be affected by local or weather conditions. If echo sounders and automatic tide gauges are to be used the survey is outside the scope of this book and recourse must be had to professional hydrographic advice and literature. If possible, a lead line should be of plaited cord rather than 'laid' rope which is less easy to handle when wet; it should be marked at metre, foot, or fathom intervals with easily recognized metal or coloured cloth tags, it should be long enough for the maximum depths to be encountered and the weight at the end should be about 5 lb (2.5 kg) so that the moment when it touches the bottom can easily be recognized. Normally the boat should be stopped for each sounding unless the observer finds that with practice he can cast the lead sufficiently far ahead for the line to be vertical as the boat passes over the spot. In general he will find it easier to concentrate on the sounding and have someone else recording and observing any fixes, with a third person engaged solely in managing the boat; but if necessary, and working more slowly, two people can manage provided that the weather and the current conditions are not too difficult. When plotting the results the soundings are first listed with their times (if in tidal waters) and are then corrected to some suitable datum. Their positions are then determined by adjustment of the lines of the soundings between the framework points and the resections along them. The corrected depths are then written alongside these positions. The adjustment is virtually the same as that of using a stationary and a moving barometer.

Other surveys

Submarine surveys

Even in underwater surveys that are carried out from the water surface the conditions largely dictate the techniques to be used. If the surveys are carried out wholly underwater then this must be even more true. Usually angle observations are almost impossible and the survey must be based on distance measurement. Expensive and sophisticated sonar systems are available but we are not concerned with these; a simple chain survey is possible with a graduated tape or line. This will require two people to hold the ends of the tape unless the area being surveyed is very small, and it must be graduated with marks and figures large enough to be read underwater and through a mask, or possibly with marks that are recognizable by touch. Where the area is relatively flat, such as an ancient archaeological site or some features of the natural sea bottom, then the techniques of ordinary chain survey will apply, with any slope of the tape being ignored. Where the feature is a recent wreck, which is likely still to have considerable heights above the bottom, it will be worthwhile to obtain design or other drawings of it, so that only a few measurements will be required to establish its identity and how it is lying; all the rest can then be filled in from the drawings. In this case, and especially where no such drawings are available, extensive reconnaissance and planning, including the making of a rough model, will also be worthwhile before starting the actual measurements so as to ensure that these are made according to a programme that will build up a series of fully measured interlocking triangles. These must create a sound framework in three dimensions from which detailed measurements can then be taken to determine the shape and size of the object being mapped. All of the above assumes a relatively small and inexpensive expedition. If more elaborate techniques are being used and there are large funds available then advantage should be taken of the ability to float a television camera above the site and to use normal aerial survey techniques. This involves the survey and identification of enough fixed features on the site to scale and orient successive overlapping pictures taken from unknown positions above it,

96 OTHER CONSIDERATIONS

followed by reproduction of the pictures in a form which can be used in a three-dimensional plotting machine. In this case the funds available would make it possible to employ photogrammetric consultants and it would obviously be worthwhile to do so.

Cave and tunnel surveys

Work underground is also carried out under some difficulty. The main problems are the need for artificial light, and there may also be running and dripping water; however, there is no air movement or refraction gradient. The scientist underground is surveying a surface from the inside instead of the outside; this makes angle measurement the easiest technique since lines of sight are relatively clear and direct distance measurement – to a cave roof for example – is not possible. At least parts of caves, as well as most tunnels, tend to be long and narrow, so the best form of framework is usually a traverse. Tunnels are normally reasonably straight or gently curved and traverse legs can therefore be fairly long, but in crooked and narrow cave passages care will have to be taken with the centring of the instrument and the target unless a magnetic compass is used to measure bearings instead of angles. For simple surveys a compass will be best, provided that there are no changes in the magnetic declination and that all metal equipment, including possibly protective helmets, is kept well out of the way. As in all traverse work some sort of closure onto itself or on to some outside framework is necessary, otherwise the traverse must be run in both directions using different stations if possible. For accurate work with a theodolite three tripods with interchangeable centre base plates should be used for the instrument and targets. In narrow tunnels these may have to be fixed to brackets attached to the tunnel wall or to beams wedged across it instead of to tripods for which there may not be sufficient head room. In very narrow tunnel or cave passages where there may not be room to get past the instrument between forward and back pointings, it may be necessary to have two observers. There are likely to be considerable changes of level in a cave passage and these must be observed with an Abney level for simple surveys or with a theodolite for more advanced ones. If the entrances are nearly horizontal, initial and closing bearings and positions should be obtained from existing maps or survey frameworks or by an exterior survey by theodolite, compass, or plane table. Where the entrances are vertical shafts, transfer of exterior positions and directions underground is achieved by suspending two plumb lines as far apart as the width of the shaft will allow. The lines joining the tops and bottoms of these will be parallel provided that the plumb bobs are heavy and are kept in buckets of oil or water to damp down their oscillations and that there are no air currents.

The traverse stations must be permanently marked with chiselled crosses and painted triangles or circles, and the detailed local survey is then based on these by intersection from them of points on the sides and roof of a large cavern and by measuring cross sections at different angles with a tape or extending graduated staff in a tunnel or narrow cave passage.† The results can then be plotted as a contour survey or, better, as a traverse in plan with elevations and cross sections along it or even by a three-dimensional transparent model. The main difficulty, apart from providing sufficient light for the observations, is likely to lie in identifying the points to be intersected from successive traverse stations; it may be possible to fire darts or arrows with coloured flags attached into the roof and sides if these are soft enough, or to take coloured photographs beforehand and to annotate them with the natural features – stalactites, coloured patches of rock, etc. – being intersected. It may be possible to carry out the whole survey by photogrammetry using a phototheodolite which is a calibrated camera whose axis can be pointed along a recorded direction. Rounds of photographs are taken with this at all traverse stations and they can then be used either in a three-dimensional plotting machine or more simply to intersect the common points seen in them. One advantage of this is that difficulties of identification are very much reduced since the identity of each intersected feature can be confirmed by looking at both viewpoints simultaneously. The other great advantage, of course, which makes photogrammetry worthwhile in difficult conditions, is that the time required for observation is reduced to a minimum and the detailed working out of the results is then carried out in comfort and not under pressure. It is a major axiom that the more difficult the observing conditions and the less time available for them due to harsh conditions, weather, or danger, the more effort should be spent beforehand in drawing up the observation programme and in designing or borrowing instruments and equipment which will save time and

†Another technique in tunnels, which was used in the Dinorwick storage scheme, was to illuminate a cross section of the tunnel by a specially made lamp giving a flat vertical beam and to photograph the illuminated cross section from a certain distance with a calibrated camera (see I. Waite and G. A. Murray *Surveying for Dinorwick pumped storage scheme* Paper B.2 Survey and Mapping 81 R.I.C.S. 1981.

OTHER CONSIDERATIONS

mental effort and the likelihood of mistakes in the field, and make it possible to do as much as possible of the detailed working out of the results afterwards in the peace and comfort of an office or laboratory (see p. 70).

Measuring the height of individual objects

Where an object is on reasonably flat ground, and the bottom, i.e. the point vertically below the top, is accessible and visible, there is no problem; the horizontal distance between the observer and the bottom is measured directly with a tape, and the vertical angle to the top is measured from the observer's end of this distance. The height will then be the distance multiplied by the tangent of this angle plus the height of the observer's eye above the ground. For rough work, if no direct vertical angle measurement is possible, the height can be measured by putting one's eye near the ground and lining up the top of a vertical staff of known length with the top of the object and measuring the distance to the bottom of the staff as well as to the bottom of the object. The height of the object will then be the height of the staff multiplied by the distance between the observer's eye and the foot of the object divided by the distance from his eye to the foot of the staff Fig. 4.5(a). Complications arise when the bottom of the object is visible but inaccessible or the observations have to be made from sloping ground. In this case the vertical angles to both the top and the bottom of the object must be observed from two observation points in line with it, and the length and slope of the straight line joining these must be measured. The result can then be obtained by drawing the figure out to scale or be calculated by simple trigonometry as shown in Fig. 4.5(b). If the bottom of the object is not visible from a distance, there would seem to be no substitute for climbing to the top and measuring the height directly with a plumb line whose length is then measured against a tape, or taking readings at the top and the bottom with an aneroid barometer. One examination candidate, when asked how he would use an aneroid barometer to measure the height of a tower, proposed suspending it from a string and then measuring the length of the string, but this is not a solution which is recommended.

OTHER CONSIDERATIONS 99

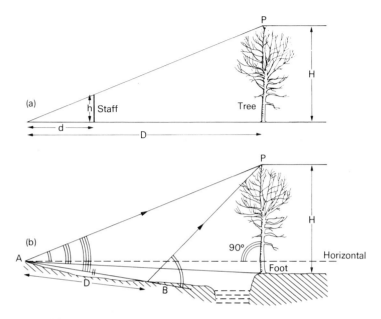

Fig. 4.5(a) Measuring the height of an accessible object on level ground
H/D = h/d.

Fig. 4.5(b) Measuring the height of an object with an inaccessible foot on unlevel ground. Measure angles of triangle ABP and side AB, and angle PA horizontal and also angle PA foot. Then draw out to scale.

PART II
Advanced field surveying

5. ANGLE AND DISTANCE MEASUREMENT

Introduction

The simple forms of survey which any competent scientist should be able to carry out with equipment that is either relatively cheap to buy or easy to borrow have been described in Part I. In all cases, except that of determining heights by vertical angles, only simple arithmetic was required, and the scientist's observations of distance and angle were plotted directly on to the map sheet using either an alidade or a ruler and protractor. Except in ideal conditions over small flat areas using a steel tape, the upper limit of accuracy in planimetry was of the order of 1 part in 1000. In this chapter more sophisticated techniques are described, in particular the use of a theodolite, which brings the accuracy of a survey framework used to control more extensive detail surveys to something like 1 part in 10 000 of a distance, corresponding to 0.33′ or 20″ in angle or direction. An accuracy of this kind means that a map can be up to 5 m long before the errors in the framework (and therefore in the detail survey) need exceed the plottable error of 0.5 mm, and that the detail surveys can start anywhere within it and will all fit. The next chapter also describes how a theodolite can be used for heights and for detail survey either with or without an assistant holding a staff; once the scientist has mastered the handling of a theodolite he may well find it an advantage to use it for detail as well as framework surveys and thus save the weight of a second instrument such as a plane table.

In proceeding to the more advanced techniques of angle and distance measurement which require the use of a theodolite and steel tape and the computation of co-ordinates, the scientist is clearly moving into a part of survey technology normally occupied by the specialist land surveyor. This means that he will have to spend a good deal of his time outside his own subject, and he must think carefully about the economic justification of this. His survey results will cost more and be of more lasting value, provided that he is prepared to match the extra instrumental and computational effort with a proportionately extra effort to ensure the permanence of his framework on the ground as well as on paper. A survey framework observed to 1 part in 10 000 in planimetry and to the order of 1 cm in height is a structure of permanent value likely to be used by others — if they can find the marks — long after the observer himself has departed. Marking it properly with precise, unambiguous, and permanent monuments in the ground is therefore not only an integral part of the survey but is common sense, since the extra cost of doing this instead of relying on wooden pegs will only be a fraction of the total cost of reconnaissance, observing, adjustment, and computation — or of doing the whole job again at a later stage. Thus in the first section of this chapter we describe in some detail the principles and practice of establishing good ground marks. After this we describe the techniques and instruments required to achieve these higher accuracies, first of all in handling the two main instruments, the theodolite and the steel tape, for angle and distance measurement respectively, and then in the use of these for the more refined and expensive techniques of triangulation and traverse of which the basic principles and simple practice have already been outlined in the relevant sections of the previous chapters.

As well as attempting to show an interested field scientist how to do his own land surveying to something like professional or good technician standards, this chapter will give him an idea of what he should expect from a contract firm or local government department engaged by him or his organization to do work of this kind as a basis for his own usually less precise detailed mapping. It is a sad fact that even the most respected firms or survey departments headed by highly qualified partners or directors do not always know what is going on in the field, while the field scientist working alongside their field surveyors usually does, and this is of course particularly true when working overseas thousands of miles away from the head office. One common reason for poor work by contractors is that they have engaged temporary staff at short notice to fulfil a contract, and it is these people in particular who may be working to much lower standards and to wrong principles. Armed with this book an intelligent and observant field scientist should be able to get at least some idea whether the job is being done correctly, and whether in a year or more's time the engineers or agriculturalists implementing his proposals will in fact be able to use the topographic maps and the survey framework, or whether the latter will have completely disappeared and will have to be established all over again, or will not be accurate enough for them.

Marking the framework

The best form of precise *ground mark* is either a wooden peg for temporary use, or a buried cube or cylinder of concrete with a hole or metal stud in the top when something more permanent is required. On rock a chiselled dot surrounded by a circle or triangle can be used. In soft ground a good mark is like an iceberg: most of it is invisible, and its top should only be a few cm above ground. While the elaborate and expensive monuments or pillars established by national survey departments are obviously not required, the field scientist should establish marks designed to be as precise as his survey and to last as long as it is likely to be used. The better this is, the better and more permanent the marks should be, for they may well be of value in the future to another scientist extending or intensifying his work, or to engineers or agriculturalists actually setting out on the ground the buildings, farm layouts, or canals and roads designed on the original map. He should consider all this before deciding how much time and materials to expend on making his marks, rather than deciding too easily that wooden pegs will be adequate because they are cheap and quick; they may have disappeared when users of his survey want to set out on the ground proposals based on his survey.

The chief enemies of survey marks are small boys, treasure seekers, political objectors, and the local Public Works Department. It is worth enlisting the help of local authorities and leaders by explaining the importance of leaving the marks undisturbed, and allaying the common fear (particularly in developing countries) that the survey precedes expropriation of the land for settlement by outsiders. It should be made clear that the marks must not be moved or you may find, as at least one surveyor has, that they have been dug out and are all safely stored in a Chief's compound! Public Works Departments are in a different category and are the principal danger to survey marks established along a road which is being improved. The answer is to site them well away from the road; as already mentioned with regard to traversing this may improve the framework and reduce the number of stations. Treasure seekers are usually after metal, and where this is locally valuable marks should be made without it, using a short length of plastic pipe to make a vertical hole in the top of a concrete block into which a ranging rod can be inserted.

Concrete marks, whether with metal bolts or plastic pipes, can be *precast* or *cast in place*. The former requires previous organization to find a local contractor or authority to do it, but the result is a very hard monument which can be dug in (but also of course out again) relatively easily. The mark cast in place can be larger and more difficult to extract but requires more labour in the field; the concrete will be of poorer quality, but most of the materials for it can often be obtained from a nearby local stream bed. These should be well washed to remove any earth or organic material. Examples of both kinds of mark are shown in Fig. 5.1. Pegs, unless for very temporary use, should be of hard wood, especially of course in termite areas. If driven in nearly up to the head they are difficult to extract. In sandy country long metal rods or pipes may be used.

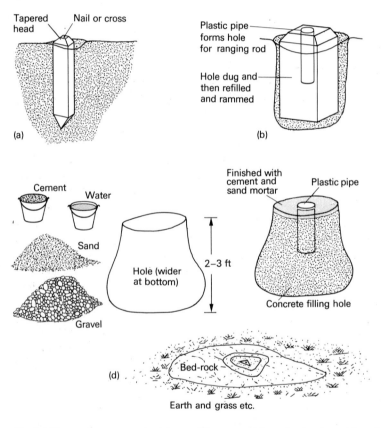

Fig. 5.1 Types of survey mark: (a) peg; (b) precast; (c) casting in place; (d) rock mark (*not* on loose boulder).

Whatever form of mark is used, unless it is required for a very short period and only by one surveyor (which must mean a relatively small project), some form of description is necessary so that it can be found again later on. This is particularly necessary in conditions where it has to be as inconspicuous as possible because of the hazards already described. In a developed area with buildings this is of course relatively easy to do by making a sketch of the area round the mark and measuring distances and bearings to almost permanent features such as house corners, bridge abutments (along a road traverse), blazed trees, etc. In a completely undeveloped area this may be almost impossible, especially in the kind of featureless earth or sand plain where irrigation schemes are so often projected. One answer in such cases, and it may be required anyway for the mapping, is to identify the point itself on air photographs, but because of the dangers of misidentifications (as described in Chapter 9) a better approach is to identify some clear and unambiguous features round it and record the bearings and distances to them. Soil and other agricultural surveyors should make sure that they obtain from the local survey department (or from an organization like the Directorate of Overseas Surveys if it holds the records) the descriptions of all survey marks already in existence in their area. These may be very elaborate and include details of witness marks put in at the same time, and even (as in the Ordnance Survey records) photographs of the actual mark and its surroundings; but for lower-order marks in featureless areas there may be nothing but the co-ordinates simply because there were no permanent features in the neighbourhood. However, where the marks are in agricultural land already in use, the local cultivators and farmers will usually know where they are and this may well be the best way to find some of them. The others can then be located by bearings and distances derived from their co-ordinates.

Angle measurement with the theodolite

In the previous chapter we described how horizontal angles can be measured and recorded by a magnetic compass to at best a quarter of a degree and to about 1 part in 1000 (about 3') by plane table in the direct graphical plotting of a long ray. Vertical angles can be measured to something like the same accuracies with an Abney level or an Indian clinometer. To achieve higher accuracies than this the scientist must use the symbol of the land surveyor – the *theodolite*.

If he has felt the limitations of the simpler techniques he will already have some idea of what this instrument must be like. The ordinary level, and possibly a telescopic alidade, will have already shown him that a magnifying telescope fitted with fine cross hairs or a graticule makes possible much greater accuracy of pointing compared with using the open sights of an ordinary alidade, clinometer, or prismatic compass. The increased precision of pointing made possible by a telescope is related directly to its magnification and resolution, and since most instrumental telescopes have a magnification of at least 20 × this precision of pointing is clearly of a different order altogether, being in the neighbourhood of 1 part in 10 000-20 000 which corresponds to seconds rather than minutes of arc. Because the pointing accuracy of a telescopic alidade is not matched by its drawing accuracy, it is hardly worth its extra weight. The thickness of a hair or graticule line in the telescope diaphragm subtends an angle of a few seconds at the eye of the observer, and with a properly designed symmetrical target the pointing accuracy can certainly be of the order of $10''$ in the lowest order of theodolite. The rest of the instrument must therefore be designed to take advantage of this and to make it possible to read the individual pointings and hence the angles between them with comparable accuracy.

The theodolite (Fig. 5.2) consists essentially of a telescope, with cross hairs or an engraved glass graticule, which can be rotated about both a horizontal and a vertical axis. A graduated circle is attached to each of these together with some method of reading each of them which includes a clamp and slow motion screw as well as verniers or micrometers. Each horizontal angle is measured between the rays to two different targets, while each vertical angle is measured to one only, the other arm of the angle being defined by the horizontal line from the instrument in the direction of the target. Although in principle the design and use of the theodolite would seem to be simple enough, in fact there are several refinements required so that the true accuracy of measurement, as opposed to the apparent precision, is comparable with the sensitivity of the telescope pointing. A number of potential errors have to be eliminated, or at least reduced to acceptable limits. For example the angles measured must be either truly horizontal or truly vertical, and this means that the theodolite must be levelled a good deal more accurately than a compass or plane table,

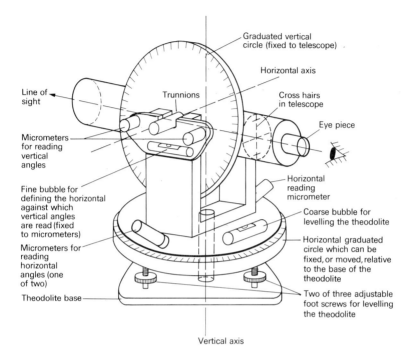

Fig. 5.2 The theodolite. Clamping and slow-motion screws are not shown.

and not just by eye. Moreover, its horizontal axis must be at right angles to the vertical axis; and the optical axis of the telescope, defined by the central cross of the graticule and the centres of its lenses, must be at right angles to the horizontal axis about which the telescope rotates. If it is not, then clearly while angles measured between horizontally sighted targets will be correct, those measured between targets at different angles of elevation or depression will not. This is the *collimation error* which is eliminated by observing on both faces, with the vertical circle first on the left-hand side and then on the right-hand side of the telescope, when this error will be equal and opposite. Since vertical angles are measured against the horizontal, this must be defined by a sensitive bubble tube, and clearly the axis of this must be adjusted so that when the vertical circle reads zero the telescope's optical axis is truly horizontal, in the same way that the telescope axis and that of the bubble tube in a level must be exactly parallel.

The circles are graduated to $10'$ or $20'$, and verniers or micrometers are used to interpolate between these graduations. In old instruments there are two of these reading devices (micrometers) for each circle which have to be read at the sides of the instrument to eliminate a further possible source of error — *eccentricity* of the circle (i.e. its centre of graduation not coinciding with its centre of rotation) which would make the measurement of an angle on one side of it only either too great or too small. In later models only one micrometer is used which combines near the eye end of the telescope images of both sides of the circle simultaneously and observes the coincidence of two diametrically opposite graduations. Some small modern low-order theodolites are now read on one side only of the circles because the makers are confident that the eccentricity is too small to have any appreciable effect, but the principle is unsound.

Having outlined the principles governing the design and use of the theodolite, in the following paragraphs we shall try to tell the field scientist what to do when faced with a tripod and a box labelled 'theodolite', perhaps some months after being given practical instruction in its use. The first step is to select a level space near the base or camp from which at least two distant sharp objects can be seen, such as a flagpole, a church spire, or a minaret or similar man-made object with both vertical and horizontal features which can be accurately intersected by the cross hairs of the instrument. The next step is to erect the tripod, and to practise doing this so that the top is level and also centred within a centimetre or two vertically over a ground mark such as a peg. This can be checked by hanging the *plumb bob* in the instrument box from the hook under the tripod head or by dropping a small object from the tripod head. The tripod feet must be trodden well in unless the surface is hard; if this is not possible (as on a floor or on concrete) a triangle of string should be fitted to stop the feet splaying out. All screws should be well tightened and the tripod head will then be level, firm, and centred approximately over the mark. The tripod head will be one of two kinds, depending on the age of the instrument. The older type has a threaded ring of large diameter with a cap screwed on, while most modern ones have a flat triangular head with a bolt running in a pivoting groove underneath which is screwed upwards into the centre of the theodolite base. Some tripods have a small pea bubble on the top which makes levelling easier. The case should now be opened, but before taking the theodolite out its position in the case should be noted carefully (there may be a picture on the lid), and if there are no letters or symbols marked on both instrument and box it will be worthwhile to add one or two, such as coloured blobs of paint or tape. More frustration, and sometimes more damage, is caused from incorrect replacement of instruments in their cases than they suffer during observation or transport. Having noted its position and taken it out, grasping it by the upper section, the theodolite is placed on the tripod and either screwed directly down on the threaded head of the tripod or screwed down onto the flat head by the bolt underneath the tripod head. In both cases the clamping screws for the lower plate should be loosened and the foot screws set in the middle of their runs, and in the latter the triangular base of the theodolite should be fitted approximately over the top of the tripod with their corners coinciding and not alternating.

The theodolite should now be roughly level and *centred* over the mark, both being within the ranges of the instrument's own adjustments for doing these exactly. Precise centring should be done first, though a final check on this is required after the precise levelling is finished. The plumb bob from the instrument box is suspended from a hook under the instrument and centred over the mark, if necessary shielding it from the wind. Common sense and the usual rules of thumb should be used in deciding how accurate this should be. With a triangulation using a vernier theodolite reading to 20' (1/10 000) and with sides 10 km long, the precision of angular measurement cannot be better than 1/10 000, and is probably worse, so a centring error of several centimetres is of no consequence. However, in a traverse along a path where the legs may only be tens of metres long, the centring should be done to a millimetre or two. In fact, in precise traversing in towns or in engineering work with possibly even shorter legs, professional surveyors use a three-tripod system with *forced centring* in which the instrument and the targets are interchangeable so that centring is accurate to even better than this. In general this refinement should not be required by scientists (except perhaps in caves — see p. 96).

The instrument should now be levelled using the three foot screws between its horizontal circle and its base. No one in their senses uses an instrument with four screws (who ever heard of a four-legged milking stool?), though some older models with four screws are sometimes seen. The theodolite is turned until the bubble on the circle cover lies parallel to two of the foot screws, and these are turned in opposite directions until it is centred. The theodolite is then turned through a right angle and the third screw is adjusted until the bubble is centred again. If it is correctly adjusted the theodolite will now be very nearly level. A further right-angle turn will bring the bubble parallel to the first two foot screws but the opposite way round, and if it is not centred it may be necessary to adjust it. A similar check over the single foot screw should be made, and if the bubble has an obvious centring error which persists in all positions then half this error should be taken up by adjusting the two capstan-headed screws at one end (as already described for the level in Chapter 3) and the other half by the foot screws. Provided it stays centred all the way round within one or two divisions the instrument can be regarded as level, and angular measurement may begin.

The upper part of the instrument, *the alidade*, is rotated with the circle on the left hand as seen from the eye end until the reading in the horizontal micrometer is near zero when pointing at the first object. (In modern theodolites the horizontal reading can be changed by a special screw.) The eyepiece should be adjusted until the cross hairs are sharp, and the objective then focused on the object. Final focusing is checked as in levelling by moving the head sideways to see if its image moves relative to that of the cross hairs, since, while both may seem to be in focus, for accurate pointing both images must be on the same plane, that of the cross hairs. The first (usually left-hand) object should now be intersected accurately by the vertical hair just above or below the horizontal hair with the final pointing made by turning the slow motion screw in the right-hand direction; it presses against a spring plunger and if the final turn is in the left-hand direction an error can be caused if this takes a second or two to act. A check on the pointing should be made with 'hands off', i.e. no contact with the instrument. The verniers or micrometers should now be read and the telescope turned on the same face, loosening the top clamp and leaving the lower one clamped, to the next object which is intersected, and the reading is taken in the same way. The objective focus may be changed but not that of the eyepiece. The telescope is then rotated on its horizontal axis so that the vertical circle is now on the right, and the process is repeated.

Reading systems on theodolites vary a good deal and it is only possible here to outline the four main different types; a handbook for the instrument used should be in the case with it (or should be obtained before going to the field if it is not) and this should be consulted. The four main types are: two verniers, two micrometers, a single double-reading micrometer, and a single-reading micrometer. In all of these there is a coarse reading mark which is seen against a small part of the circle and identifies the particular graduation from which the fraction of the division between it and the next is then measured precisely so that the final reading is in two parts: that of the graduation (which is usually in tens or twenties of minutes of arc) and the minutes and seconds to be added to it. Most scientists are familiar with *verniers* and *micrometers* for measuring small distances, and the principles of angular measurement are exactly the same since the small portion of the circle over which they operate is virtually straight. The vernier depends on a short scale placed alongside the main scale with divisions which are slightly shorter (say by one-tenth). Thus ten divisions of the vernier scale correspond to nine of the main one, and the point where the two coincide indicates, on the vernier scale, the required fraction of a division on the main scale. Not much can go wrong with the vernier itself, but it is often difficult to read and it is easy to make a gross error. In several developing countries vernier theodolites are still used by the lower grades of their government surveyors.

In a micrometer of the old two-micrometer system (Fig. 5.3) the coarse index mark indicates the approximate reading by its position between two adjacent graduations of the main scale, and the precise proportion of this division is obtained by rotating the micrometer drum (whose scale corresponds to that of one division on the circle) until two parallel hairs exactly straddle one of the main scale graduations; the precise reading is then taken from the drum against its own index mark. Clearly the drum should read either zero or the end of its scale when the coarse index mark is opposite a graduation and when the hairs are set over a graduation, and adjustments may have to be made to its *run* to ensure this, as detailed in the handbook of the instrument. In modern theodolites the micrometers are not at the sides but are read from a telescope aligned with the main one so that the observer does not have to move round the instrument to read it, which is a great convenience and increases both speed and accuracy. In the best instruments the principle of reading both sides of the circle simultaneously (to eliminate any eccentricity of the axis) is preserved by matching up two opposite graduations to set

114 ANGLE AND DISTANCE MEASUREMENT

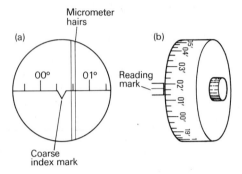

Fig. 5.3 Old two-micrometer theodolite. (a) View in microscope (circle divided to 20′). The reading is

coarse	00°	20′	
fine		01′	40″
estimation			10″
total	00°	21′	50″

(b) Micrometer drum (drum divided to 20″).

the micrometer (Fig. 5.4), but the smaller less accurate instruments are now so robust and well made that in many of them the micrometer reads only one graduation on one side of the circle. It is all the more essential in such instruments to read all angles on both faces since this will clearly help to eliminate any eccentricity effect as well as the collimation error, because on the opposite face the single micrometer will be reading on the opposite side of the circle.

Other important features of modern theodolites are their lightness and small size due to the greater precision with which their circles are made, internal focusing of the telescope which means it can be quite short and sealed against dust and fungus, and the use of glass circles instead of brass or gun metal ones which have to be read by reflected light. Glass circles can be read by transmitted light which is clearer and less liable to cause uneven illumination or *phase error* in the image of the graduation. Field scientists are usually obliged to borrow survey instruments from a university or college teaching department; they should try and obtain modern instruments, but they may have to be content with old fashioned heavier models. All teaching establishments have these, both because they are cheaper and because the principles of theodolite adjustment can be seen and

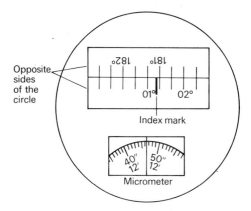

Fig. 5.4 Coincidence reading in the Wild T_2 theodolite. The reading is

	coarse	01° 00'	
	fine	12'	46.5"
	total	01° 12'	46.5"

carried out on them easily while in a modern theodolite it is all under cover, the optical train is complex, and only a very experienced and handy land surveyor would dare to try and adjust its micrometers. Thus these older instruments have advantages and they are quite accurate enough for most purposes, but they will require more effort (or assistants) or carry them and they are slower to use. The main advantages of modern instruments are seen in rugged country where labour is in short supply and observing sites (like rocky hill tops) are difficult to reach and awkward or even dangerous to occupy. When costing a long expedition the time saved by using a modern instrument may even justify buying one.

Having now set up, centred, levelled, and read the instrument on both faces and obtained two values for the horizontal angle between the two distant objects, the surveyor will find that these are not the same. If the difference is only about a tenth of the whole vernier or micrometer scale (the spacing between two graduations on the circle) then the instrument is in adequate adjustment, but if the difference is much greater, then the collimation requires adjustment and this can easily be done even on modern instruments. The theodolite is turned again to the first object on face left, read, and turned to the second object. The micrometers are now set to read so that the resulting angle would be equal to the mean angle obtained from both face-left and face-right readings, and the telescope is turned by the slow-motion screw so that the micrometer hairs now exactly intersect the graduation or the vernier reads this value. Clearly the cross hairs will now no longer intersect the target, and they should be made to do this by moving the diaphragm which carries them inside the telescope. This is held by four capstan-headed screws at right angles; the top and bottom screws are loosened and the two side screws are turned in opposite directions until the vertical cross hair intersects the target. The top and bottom screws are tightened and a check measurement of the angle is made on the other face; it should clearly now give almost the same value and the collimation error has been removed.

It is axiomatic that in all except detail shots theodolite angles must be measured on both faces; this is not only for greater accuracy but is an insurance against misreading and introducing gross errors or blunders, either in reading or in computation, since in both cases different values will be read and averaged and subtracted in degrees, minutes, and seconds. It is also advisable in observations for the main framework to repeat the horizontal angles on a different *zero*, i.e. starting with the reading for the first object (the *reference object* or RO) on a quite different part of the circle. If two zeros are used they should be 90° apart, if four are used they should be 45° apart, and so on; for very precise work geodesists observe up to 16 zeros spaced equally round the circle and on different days, but obviously this is not required in the sort of frameworks considered here. Two zeros is the minimum, but more may be advisable at a particular station for two reasons. The first is that if it is difficult or time wasting to reach a station, one wants to be sure of not having to re-observe there. The second case is if the station is at one of the apices (corners) of a triangle where the angle is small, i.e. below 30°. This angle approaches what is called a *distance angle* because the errors in it cause a much greater effect on its distance from the other two apices of the triangle than will occur at them with their angles approaching a right angle, for obvious reasons. It may therefore be well worthwhile to observe this relatively small angle to a higher degree of accuracy than the other two, and in adjusting the misclosure of the triangle to give it greater weight and allot it only (say) a sixth of the triangle misclosure instead of the usual third. Unless the stations are very close (as in a traverse with short legs) the time taken to observe an extra zero or two once the instrument is set up and levelled is quite a small fraction of the time taken up by moving from one station to the next and is obviously well worthwhile. Moreover, all observations at a station should be averaged and reduced to angles and compared either by the observer himself or by a numerate and trustworthy booker, if he is lucky enough to have one, before the instrument is taken down. On each round of angles where more than two objects are being observed, there should be a final shot back at the RO (reference object, i.e. the first object sighted) to check that the instrument has not moved during the round, and it is customary to observe in opposite directions on the two faces, i.e. face right swing right and face left swing left. This is done to eliminate the residual effects of any axis strain or slipping. In one case, when observing from a trig pillar heated unevenly by a setting sun I found large closing differences on the RO and these appeared progressively at each object in the two sets of observations on each face, but because the swings were opposite on each face this movement of the pillar as one side cooled down cancelled out and the results of the angles on two zeros were satisfactory as were the triangular misclosures which are of course the final test.

Recording

Various styles of recording angle observations are used, and the observer can devise his own or adopt that of some survey organization which can provide printed booking forms. These follow the general principles that each station starts on a fresh page, and each pair of pointings (face left and face right) to an object has a line to itself, starting with the RO, with successive columns showing the object, the readings on face left and face right, the mean of these readings, the reduction to the RO, i.e. zero for this and the angles measured clockwise from it for the other objects and then finally a column for remarks, e.g. any doubt about whether the object was correctly seen. Other spaces on the form are provided for the names or numbers of the project, the station, the observer and recorder or booker, and the instrument used. It is very important to note all relevant details of the station, particularly of course if the actual centre of the mark was not occupied and a satellite station had to be used; in this case a complete diagram showing the relationship of this and the station mark to the rays observed should always be drawn on the field sheet. An important principle in surveying is that all original observations should be shown, even if a mistake has been made and discovered; the wrong figure must be cancelled by a diagonal stroke and not rubbed or scratched out. The deletion and replacement of an observation suggests that it may not have agreed with the computed results, i.e. there had been *cooking* or *fudging* which is the surveyor's cardinal and unforgivable sin. It pays to use loose-leaf forms rather than books (except for levelling) because these can be extracted and filed at base once they are complete and not put at risk in the field on successive days when they might get lost or wet. Some systems have carbon copies with a sensitive duplicate for each original; these can then be retained if the originals are sent back for computing or simply as an insurance. Separate carbon paper is unusable on the windy hill tops where trigonometric surveyors spend much of their time.

Having practised the whole procedure of unpacking, setting up, observing, and recording by himself in the quiet of his base or camp, the surveyor is now ready to visit his first station; it is a great mistake to go there before being reasonably at home with the instrument and sure that it is in adjustment. It should also be noted that while levels can be carried on their tripods over the shoulder, a theodolite with its extra weight and height and sensitive axes must never be carried more than a few yards on its tripod, and then only in an upright position.

Accurate distance measurement

In Chapter 2 a warning was given about the actual complexity of accurate distance measurement compared with its apparent simplicity if the sort of accuracies required for a major framework survey are to be achieved over long distances and rough terrain. This is particularly true if the distances required are for bases scaling a triangulation; there will only be two of them and it is worth taking a little extra trouble to make them accurate enough so that in adjusting and checking the triangulation between them they can be regarded as errorless. During the reconnaissance an eye should be kept open for suitable sites; a good base should be reasonably level but with elevated ends so that they not only see each other but at least two points of the main framework as nearly as possible opposite on each side on the middle of the base, forming a diamond of which it is the median line. Alternatively it may form the shortest side of a quadrilateral of which a main triangle side is the longest so that the accurate short length of the base is transferred to a side of the framework with the least possible loss of accuracy. With traversing there is less choice of terrain and the emphasis is more on avoiding gross or systematic errors which may require the complete remeasurement of a long traverse because it has not closed properly.

The commonest method of distance measurement will be by steel tape, and the surveyor will have to decide whether to measure all distances in both directions as a check, or save time by measuring them only once and rely on his triangulation or traverse closures as a check. For the field scientist the former is without doubt the best answer because being inexperienced he may well make mistakes, and if the triangulation chain or traverses are long this is expensive. If possible two different tapes of different lengths and with different units should be used, since this provides a virtually foolproof check on gross errors in reading the tapes and in counting the number of tape lengths. In any observations a good surveyor carries out checks as he proceeds so that when he arrives at a major independent check (as in closing a triangulation or a traverse or levelling circuit) he is 99 per cent certain that it will be correct. Moreover, in the tropics angular observation to opaque targets (i.e. not lights or helios) is only possible in the early morning or late afternoon, even in cloudy conditions and under trees, so there is plenty of time in the middle of the day for taping (or chaining as surveyors tend to call it even when a tape is used).

A major decision required in accurate distance measurement by tape is whether to tape along the ground or in the air in catenary. In any except very flat ground the former requires more clearing, and in the tropics may require a good deal of cutting through termite mounds and more labourers to hold the tape level and stop it sagging over hollows; this soon becomes expensive, tedious, and slow. Catenary taping requires calm conditions, the use of a little extra equipment, and some extra computation, but it is faster, more accurate, and requires fewer assistants. It is also more intellectually challenging, and the graduate of whatever scientific discipline will find it more worthwhile if doing the work himself, unless the ground is so flat and open that it is prevented by windy conditions and is also obviously unnecessary.

In both methods the tape must be laid successively along the straight line joining the two stations at the ends of the line, and this is done by aligning the forward labourer dragging the front end of the tape on the distant mark from the back end, using a theodolite if the ground is very uneven. In ground taping, when the whole length of the tape has been extended, with the zero on the starting mark at the back, an arrow or peg is put in just short of the end of the tape; then, with the man at the back holding the zero against the mark of the starting point and the forward man pulling the tape taut, the surveyor reads the tape against the peg or arrow. This is a better practice than putting the arrow in at the exact end mark of the tape; as in all survey work it is better to put a mark in first and then measure to it and record the measurement than to try and put it against an exact value; the results are likely to be more accurate because the observations are more objective. While he is measuring at the front the assistant at the back should be shouting that he is maintaining the zero accurately against the mark; and if the ground is uneven men will be required to hold the tape level and straight. The distance should be measured at least twice with the tape slackened in between, and for accurate work a spring balance should be used at the forward end under the surveyor's eye; the usual tension is 10 kg or 22 lb. The slope of the tape length or *bay* should then be measured; if it is under 5° an Abney level is quite accurate enough, with measurements taken from both ends. For very accurate work, or in extreme temperatures, the temperature of the tape should also be taken by laying a thermometer (in a metal case) on the ground, although for most purposes an average can be assumed for every hour's taping. The tape is then taken on to the next bay of the base or traverse leg, and the procedure is repeated until the end of the base or the next traverse station is reached when there will normally be an odd length to be measured; here special care should be taken to see that no gross error of reading units occurs.

In *catenary* taping the tape is suspended under a known tension between two or more supports (in a curve known as a catenary) so that none of it touches the ground. A tape over 50 m long will require at least one intermediate support, and when possible these should be aligned vertically as well as horizontally between the two ends by the surveyor sighting back from the front end. Where this is not possible, i.e. over a convex slope, the slopes of the different suspended sections of the tape must all be measured and allowed for, and over a valley the full length of the tape may be unsupported. At the two ends the tripod against which the tape end was measured, or the ground mark to which a measurement is taken, must be guaranteed not to move between measurements taken to it from behind and those taken forward from it. This can be achieved either by using fairly heavy and well-set tripods and measuring to and from a mark on their heads, or by pulling each end of the tape with a ranging rod as a lever and plumbing the two ends down to pegs set on the ground. The former is preferable because it is quicker and it eliminates the errors caused by the use of a plumb line. One end of the tape can be held against the axis of the theodolite, which is used alternately at the two ends of the tape, and is also used to measure forward and back slopes alternately. If two separate tripods are used they should be fairly tall and rigid and have well-shod pointed feet for pressing firmly into the ground, and men will be required to bring the back one forward at a run. In this case an Abney level can be used for measuring the slopes. Temperature measurements are taken with the metal-encased thermometer in the same situation as the tape – i.e. suspended in the air but with the sun shining on it if the tape is also in the sun.

Electromagnetic distance measurement

Electromagnetic distance measurement (EDM) is the modern answer to the tedium of distance measurement in traversing, but as the instruments cost several thousand pounds each and are still liable to go wrong, especially in hot humid conditions, they still tend to be used only by land surveyors or engineers and not by other field scientists. There are now so many different makes and models available that detailed descriptions would not be justified especially as new models are appearing frequently. However, one or two points can be made. The first is that any competent scientist (especially one with practical experience of electrical equipment) should be able to learn how to operate one of the short-range instruments after a few hours' instruction if he is supplied with a handbook and a few spares. However, he will be adding to his responsibilities and insurance costs, and he will also require a spare car battery and a charging plant if his base does not have mains electricity and someone will have to look after these. For long ranges (over 2–3 km) two measuring sets using microwaves or a very expensive single set using a laser are required, and this is clearly beyond the scope of the ordinary field scientist. However, for extensive traversing over undulating country, and especially if reasonably long traverse legs (around 1 km) are possible, a single short-range EDM instrument and reflectors will save a great deal of time and possible error. Each measurement of the length of a whole traverse leg takes only a few minutes and is accurate to something of the order of 1 part in 100 000 over such distances; no clearing is required beyond that needed to obtain the theodolite line of sight and to ensure that the two ends of the distance are intervisible, and the distance can be measured from the same tripod set-up as the angle. In deciding whether or not to try and borrow this equipment the amount of traversing required, the ruggedness of the area, the lengths of the traverse legs, and the competence of the scientist to look after an electronic instrument and its electrical paraphernalia should all be considered. In any case steel tapes and their equipment must also be taken as a reserve.

Most short-range EDM instruments give a direct distance reading, but some of the older ones the field scientist may be lent require a series of readings to give the centimetres, metres, and hundreds of metres, and he may have to know his distance to the nearest kilometre. Some require the setting of a 'null meter' before the correct reading is obtained, and for the greatest accuracy the air temperature may have to be taken although some of the latest models are compensated for this. Some of the older models can give false readings if the beam is interrupted during measurement, and the scientist should check whether the instrument is proof against this. All models operate by timing the double path of a modulated high-frequency signal out to a reflector or bank of reflectors at the other end of the distance and back; it is the attenuation of the returning signal that restricts their range and this can be increased if a laser is used. The cheaper and lighter models use an infrared instead of an optical signal so no reflection can actually be seen, but the coarse alignment of the beam on the target, which is done using a telescope attached to the instrument, can be made more accurate by watching a meter showing the strength of the returned signal which must be brought to the maximum. A check against gross reading error can be made by moving the instrument forward or backward a metre or two and taking a new reading which is checked by taping the distance between the two set-ups. A simpler technique is to measure every distance both forward and backward; although this requires twice as many set-ups as measuring each distance once, only those who have found that they virtually never misread the instrument can afford to neglect this precaution. A danger for field scientists is that the virtually error-free length results may tempt them to increase the accuracy of their angular measurements or of the whole survey beyond what is required; they should not forget that for them the main advantages of EDM are its speed and saving of labour; it measures distances up to a kilometre or two in minutes when to tape them would take an hour or two and might require more clearing and labourers to hold the tape straight.

Computing

The following *corrections* have to be applied to reduce the actual distances measured by tape to horizontal distances at sea level, and if necessary to transfer them to a national grid on which the co-ordinates of previous work may have been computed. The first three are instrumental corrections, the next two relate to the way in which the tape is used, and the last two are required to bring the results into conformity with adjacent surveys. The corrections are (1) standard, (2) tension, (3) temperature, (4) sag, (5) slope, (6) altitude, and (7) scale correction. The first three are interrelated in that the *standard* length of a tape is usually defined for a certain tension and temperature. These values should be available from the makers or owners of the tape, but often they are not if one is borrowing an old tape and the best course then is to check it against an official base line; most survey departments have one laid out in their buildings which can be used. The British National Physical Laboratory will check tapes, but this is expensive and unnecessarily accurate for the sort of work undertaken by field scientists. If no such local base can be located, the nominal length of the tape will have to be accepted, but in this case a careful check should be made the whole way along it to see that the units and their numbering are continuous all the way and that there are no false joins; it is quite common for these to be made by an overlap without introducing a 'splint' to ensure that there is no change in length, and this will shorten the tape by the length of the overlap. No correction for tension should be necessary if the standard tension (usually about 10 kgf or 22 lbf) is applied using a spring balance.

The *temperature* need only be recorded in extreme conditions; the coefficient of expansion of steel is 0.000006 per degree Fahrenheit (0.00001 per degree Celsius) and so in a temperature of 100°F a 100 m tape standardized at 60 °F (a common figure) will have a correction of +24 mm, which is not negligible but is still only about 1 part in 4000 of the tape length. As the tape expands the distance measured appears to be shorter, so this correction has the same sign as the temperature difference. Clearly in temperate conditions it can be ignored, but in the tropics in sunlight a black tape lying on the ground may reach quite a high temperature so that this correction cannot be neglected.

The *sag* correction varies as the cube of the suspended length; it also depends on the elasticity and the cross section of the tape and on its weight per unit length, all of which are difficult to determine. The best answer, since the correction can be appreciable and is always negative and thus systematic, is to measure it when it is at its greatest and apply it under conditions when it is least — a good principle in any scientific work. Therefore in order to measure it as long a section of the tape as possible — preferably its whole length — is suspended between two elevated tripods or other high supports over a stretch of level ground and its ends are accurately transferred to the ground by a plumb line. The distance between these two marks is then measured accurately with a tape along the ground, and the standard tension is of course used in both cases. The difference is the sag correction, and we have that $s = KL^3$ or $K = s/L^3$ where L is the suspended length, $L - s$ is the true distance, and K is an unknown constant which can be calculated from the values of s and L. Clearly, if the tape is now used with one or more intermediate supports the total sag correction (and any error in it) will be very much less; for example with the length L suspended in two equal halves it will be $2K (\frac{1}{2})^3 L^3$ or $\frac{1}{4}KL^3$. A 100 m steep tape 12 mm wide suspended in two halves has a total sag correction of about 60 mm (or over 1 part in 2000), and this will of course always be negative all the way along the line.

For all distance measurement (including EDM) a *slope* correction must be applied to each measured length (or section of it if the slope is not the same all the way along). The correction is the difference between the measured length and the cosine of the slope angles times that length, i.e. $L(1 - \cos A)$ if A is the slope angle. It can be quite large; for instance with a 5° slope the cosine is 0.996, so the correction is 40 cm over a 100 m or 1 part in 250 which cannot be ignored. It is also always negative and is thus systematic all the way along the line. The *altitude* correction is obtained by dividing the height above or below sea level by the earth's radius, which is approximately 6000 km, so this is generally fairly small. However, at 10 000 ft (3000 m) it would be 1 part in 2000 which is not negligible. It is also negative unless the line is below sea level when it becomes positive. The *scale factor* must be obtained from the local department, and normally an average value can be used for the whole project area. It can be either positive or negative, but seldom exceeds 1 part in 2000. This is only required if the new survey has to be correlated with an existing one which has its co-ordinates on a specially projected grid, usually the national grid of the country. It should be noted that in the United Kingdom the national grid *scale correction* can be of this order. Scientists should also note that in a mountainous area the altitude and grid scale corrections may both be negative and together amount to as much as 1/1000, as for example near Kabul in Afghanistan which is at one side of the UTM grid zone and at a height of 10 000 ft.

6. MORE ADVANCED SURVEYS – PROCEDURES

Triangulation with a theodolite

In Chapters 2 and 3 we described how to establish a framework with simple instruments and procedures, and in Chapter 5 we described how and where to put in precise, unambiguous, and reasonably permanent ground marks and how to achieve reasonably precise measurement of horizontal and vertical angles and of horizontal distances. In the next three sections we describe how, having mastered these techniques, the scientist can deploy them to establish an accurate and reliable horizontal and vertical framework for controlling detail surveys over an area comprising several map sheets. On pp. 144-8 we describe how a theodolite can be used for various rapid measurements of distance by exploiting its ability to measure small (distance) angles accurately, and how in the right circumstances these techniques can be used for the surveying of detail from a single convenient elevated viewpoint. The computation involved is described in the final section.

Reconnaissance

All that was said in Chapter 2 applies when planning a more accurate triangulation using a theodolite, and this is particularly true of the reconnaissance. Having decided from a quick preliminary reconnaissance that enough stations with good all-round visibility exist to make triangulation the best technique, the next step must be a detailed reconnaissance of the whole area. At the end of this the scientist should have completed the following:

(a) A sketch map showing the approximate positions of all the points and the rays which go through between them; normally this would be on one sheet not more than about 30 in × 20 in (75 cm × 50 cm). This is known as a *trig diagram*.
(b) All the stations marked by permanent monuments as described in Chapter 5 and all lines of sight cleared of vegetation.
(c) Either permanent beacons or sockets for temporary beacons (e.g. ranging rods with flags) established at each station.
(d) The selection of two *bases*, i.e. triangle sides shorter than

MORE ADVANCED SURVEYS – PROCEDURES

average but with reasonably level surfaces so that their lengths can be measured by tape; these bases should be as far apart as possible. Alternatively these may be sides of an existing triangulation bordering the area. If not, their *azimuth* or true bearing must be established by astronomical observations, or approximately by compass bearing or an existing map.

This work should be done thoroughly even if it takes several days, and it pays every time to have everything prepared so that the observing can be done as smoothly and quickly as possible. When observing with a theodolite, especially if one is not very experienced, one is surprisingly nervous and on edge, and any delay (like a line not properly cleared or a station wrongly sited) can be very upsetting and can have consequential effects on the programme – for example in requiring reobservation of previous rays directed to the station. The reconnaissance and marking is often done in two parts: the reconnaissance and selection first, followed by a retracing of steps over the whole area with a final check and putting in the permanent monuments and beacons, although with an experienced labour force some of the marking can be left to them on the simpler sites.

In this more advanced surveying the sketch map will be compiled by the methods described in Chapters 2 and 3, and the more accurate it is the more use it will be. It can be particularly helpful at a station for finding another one which is difficult to see by turning off the angle shown on the map from one station which has been identified and searching a small area with the theodolite telescope instead of having to scan a wider one with field glasses. If the area is suitable for triangulation it must almost certainly be suitable for plane tabling, and this is by far the best method to use for the reconnaissance; a compass survey is less accurate and is a second best alternative. In the reconnaissance map shape matters more than scale. This can be obtained accurately from two existing intervisible framework points if these exist in the area, or if they do not then an initial triangle side length can be obtained accurately enough by such methods as milometer readings along the road between its ends or by observing a subtense base between two hill tops (see p. 147). In a very open and rugged area a large number of stations may be intervisible and all the possible connections should be shown on the reconnaissance map, but it is a mistake to plan to observe all these since they will not add appreciably to the sort of accuracy required by the scientist and they will expand the observation programme unnecessarily and make computation longer and more complicated. There is no point in the scientist becoming involved in the field in the complicated adjustments and least square error solutions which require the services of professional surveyors and computers. If visibility is possible right across the area then the final observation programme should be broken down into one or two large triangles or quadrilaterals enclosing the area and these in turn should be broken down by similar figures (see Fig. 6.1).

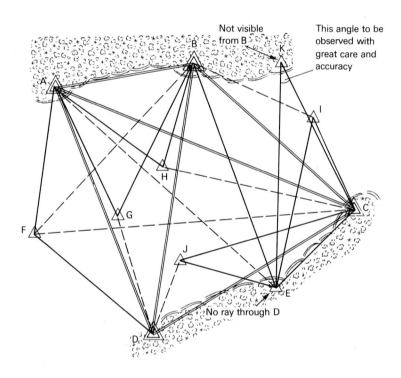

Fig. 6.1 Observing diagram for triangulation: ▲, major trigonometric points (A, B, C, D) on valley sides; △, minor trig points (E. F. G. H. I, J) in the valley; △—△, rays to be observed; △ - △, possible rays not to be observed. Note: ABCD form a fully observed strong quadrilateral. All the other points are fixed by fully observed triangles (one for each) giving the strongest fix. Because of the distance of K from CE and the small angle at K, this must be observed with extra accuracy. Bases not shown.

As long as all the angles in these are observed the resulting framework will be strong, well checked, and accurate to the precision of the theodolite or other angle-measuring instrument used. If the theodolite is not precise enough for the accuracy required, it will be more economical in both observing and computing time to observe the angles at each station on several zeros than to add a large number of redundant rays and angles all observed to the same low level; this will be particularly true for the larger figures and for the smaller angles.

The density of framework stations is determined by the scale of the proposed map and the technique to be used for the detailed survey. If this is accurate to about 1 part in 100 then the framework points should not be more than 10 cm apart at the very most. However, if detailed mapping is to be by plane table, then the positioning of the points and the areas from which they can be seen are more important than their spacing. and they can be up to 30 or 40 cm apart on the map scale, so long as in every part of the area in between at least three and preferably four of them can be seen from any one point. They may in fact even be outside the area being mapped, as for example where a flat plain or valley to be mapped is surrounded by hills whose tops will form perfect trig sites for use by a plane tabler on the plain below. It may even be worthwhile to reduce the map scale so as to include these on it. He will be able to see several of them, and fix himself by resection with only a few points actually fixed on the plain such as on minor hills or tall manmade features like church or mosque towers or ginnery or pump-house chimneys. The two selected bases should be at opposite sides of the area; each must have intervisible ends which can be occupied and from which other points of the main framework can be seen, and they should be not less than a quarter of the length of the average main triangle sides (not including the sides of any really large figures embracing the whole area).

Beacons

The permanent monuments have already been described but if the surveyor is to carry through his observing programme quickly and accurately he must also first put up suitable *beacons* or targets to which the rays can be directed. Each beacon consists of two features: a coarse part by which it is found by eye, or with binoculars, or by observing on a pre-determined direction with the theodolite, and a fine part which is used for actual intersection. In very precise work land surveyors use *lights*, i.e. either narrow-beam signalling lamps or, in an area without cloud, *helios* reflecting the sun; but both require experienced assistants to direct them in the right direction at the right time and thus a large and elaborate party which is only justified when the precision and extent of the framework are much greater than the field scientist would be observing himself. Therefore *opaque* beacons should normally suffice, with the fine part appearing to be about 1 mm thick in the field of view of the telescope so that it is about ten times the thickness of the cross hair. For distances up to 2 or 3 km a striped *ranging rod* is just right, with the coarse mark in the form of a flag whose movement helps it to be seen. These would normally be temporary and be put out each day; if so the monuments must include vertical sockets into which the flags fit so that they are upright and the surveyor can be sure that they have been accurately centred. Over 2 km and certainly over 5 km it will be necessary to erect a more permanent and larger three-legged (preferably four-legged) structure and fill in between the legs either with brush (if on a hilltop) or brightly coloured (preferably fluorescent) cloth, especially where the beacon may have to be seen against a green or brown background as when viewed from a higher altitude. A quadripod is better than a tripod because its profile is symmetrical from all points of view, while that of a tripod is not. At long ranges or in poor visibility (e.g. tropical shimmer) the narrowest part of the structure may not be seen, and if it is not symmetrical the ray will not be observed to its centre, which is of course vertically above the station monument, and so an angular error is produced. All cloth should be slit both to reduce the effect of wind and to make it less attractive to local people. As far as possible materials like cloth, wire, and rope should be obtained before proceeding to the area, and fluorescent cloth should be obtained before proceeding overseas. Types of beacons are shown in Fig. 6.2.

Cairns

On rocky hill tops a cairn is a very cheap and satisfactory beacon, but it must be built properly. The materials should be rocks which are rectangular or wedge shaped, with a length twice the breadth and a breadth at least twice the thickness, and it pays to spend some time assembling a stock of these rather than trying to use near cubic or round rocks. The wedges are then laid in a series of rings one above the other with the thin and narrow ends towards the centre so that their upper surfaces always slope inwards towards each other and the cairn cannot collapse. It should be as thick as it is high since it is the

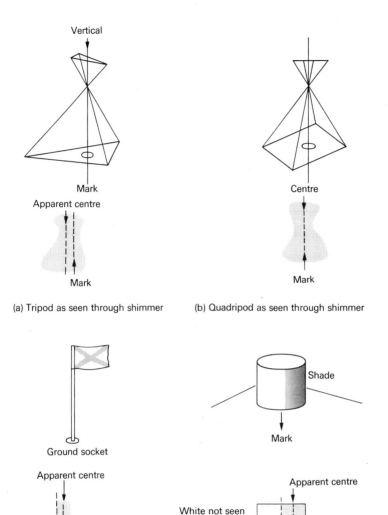

Fig. 6.2 Types of beacon and phase effect (a) tripod; (b) quadripod; (c) flag on pole; (d) white drum on pillar.

least dimension which determines whether it can be seen at a distance. Unless all the rays to it will be over 10 km it should be built round a vertical pole (painted black to be visible against the sky) centred over the ground monument because it will be too thick for satisfactory intersection itself. Once built properly it can easily be taken apart so that observation can take place over the actual mark instead of beside it. To use *satellite stations* or set-ups nearby introduces the complications and probable sources of computing error in having to correct the observations to the station mark; and since at least half the work of building a cairn is in collecting the stones, once this is done (especially if one has a few assistants) dismantling and re-erecting it is quick and easy. In tropical desert areas snakes and scorpions often seek shelter in cairns so care should be taken when dismantling them.

Phase error (see Fig. 6.2)

As already stated, opaque beacons are less accurate than lights, and although this source of error is generally unlikely to be large enough to bother a field scientist carrying out a survey, he should be aware of the errors they can cause because he may be lucky enough to have a ten or even one second theodolite and his triangular closing errors may greatly exceed the errors of individual angles in spite of careful observation and, equally important, good centring of both beacons and instrument. This kind of discrepancy causes unnecesssary concern and should be avoided. The most common source of error from the beacons is the 'phase error' due to observing a beacon which is unevenly illuminated from one side. If the beacon is clearly seen this will cause no error, but in poor visibility the observer may centre his cross hair on either the illuminated (if against a dark background) or the shaded part of the beacon (against the sky) and thus not be observing to its true centre. Drums or petrol tins silvered or painted white are particularly liable to cause this error, as would observing to a white-painted triangulation pillar, and beacons of this kind should be painted with matt black paint or covered with coloured (not white) cloth. At short ranges the effect due to shimmer can be appreciable; with a drum or pillar 1 m in diameter illuminated from one side the apparent centre will be a quarter of the diameter (25 cm) away from the true centre, and this represents a minute of arc at 1.3 km. With longer distances the errors will be smaller and often below what is unacceptable, but it is a sound surveying principle that if the discrepancies found by a check (such as a triangular misclosure) exceed by two or three times what is expected from the range of values at each individual observation (in this case of each of the three angles) then one should seek for the cause. It sometimes happens that an error larger than expected, but within the required tolerance, is due to two large errors or blunders which have not quite cancelled each other out. Thus a field scientist who is lucky enough to borrow a one second theodolite can easily achieve triangular misclosures of about ten seconds provided that the sides are not too short and that good centring and beacon design are followed; to do this gives great satisfaction as well as the practical advantage of removing any question of hidden gross errors or blunders. This is also true in a long traverse where they are more likely to occur and much more difficult to locate and correct.

Although the scientist is unlikely to be observing to lights (i.e. signalling lamps or helios), which is the most accurate type of target, he may find himself working in the field alongside surveyors engaged by his organization to observe the angles of an accurate framework in which lights are being used. In order to save the complications of having at least three parties occupying framework points simultaneously — perhaps on different mountains several kilometres apart — some survey organizations are tempted to employ only two parties and use opaque reference objects (ROs) on their own hilltops (or another) and to measure the angles from these to one distant light only; the party showing this then moves to the next station and the angle to the new position of the light is then measured to it from the same RO. The 'observed' angle between the two distant stations is then taken as the difference between the two angles actually observed to them from the RO. This is a pernicious practice for two reasons: it degrades the accuracy of the observations by using an opaque beacon for one of the pointings in each angle, and it may actually introduce a gross error if the RO used for the second observation is not the same as that used for the first. If lights are being used in order simply to ensure visibility through haze rather than for the increased accuracy they provide, this practice is of course permissible, but the following precautions should be followed. First two ROs should be used and not one, as a check against gross error. Secondly they should be as far away as possible and at not too steep a vertical angle. Thirdly they should be artificial and not natural features whose appearance, and therefore apparent position, is unlikely to be changed by the different position of the sun at the different times at which they are observed.

Suitable targets are described on p. 137 where the problems of short leg traversing are dealt with. They must be symmetrical and evenly illuminated, and if the survey party plans from the start to adopt this practice of observing to ROs they would be well advised to take with them specially made traverse targets and tripods of the type used by engineers in precise short range work; they should not rely on finding suitable objects on the top of the hill.

Traversing with a theodolite and tape

As mentioned in Chapter 2, the land surveyor traditionally always tries to avoid traversing; it is a form of single-line survey in which all the straight-line distances have to be measured directly by any means including in the last 20 years EDM, and all the angles have to be measured at each change of direction. We have already looked at some of the advantages and disadvantages in Chapter 2, but it will be as well to summarize them here when establishing a more accurate and permanent framework. The advantages are as follows:

(i) A traverse only requires visibility over relative short distances along two directions from each station, except where two traverses join.

(ii) A traverse can follow a communication line such as a road, railway, or river so that transportation is easy, including that of the materials required for building monuments.

(iii) Its marks are located in areas likely to be developed and where they are most easily reached and used by those developing the project, instead of being on remote and possibly inaccessible hill tops. Moreover, observing conditions are likely to be more sheltered. They will be less harsh in cold climates but probably more difficult in the tropics because of heat, insects, and local curiosity.

(iv) Given good communication lines, the traverse network can be shaped to fit the requirements of the detail survey rather than being dictated by the shape of the ground, e.g. the locations of intervisible hill tops.

(v) Since some of the observations are repetitive and routine they can be delegated under supervision to technicians who can be recruited locally.

The disadvantages are as follows:

(i) Checks of angles and distances are only possible by repetition until the end of the traverse is reached, and if it fails to close on itself or on another traverse, or on a point of a higher-order framework, the whole observation of that traverse may have to be repeated.
(ii) The frequency of points along a traverse is dictated by the lengths of possible straight lines of sight, and they are often closer together than is required for controlling the detail survey, while the spacing between roads or paths forming routes for different traverses may be greater than these requirements. Thus more points may have to be built and visited than in a triangulation but the control may be less well distributed.
(iii) Monuments are more likely to be destroyed by local people or by the public works department since they are more accessible.
(iv) Positional accuracies in the middle of a long traverse are always suspect however well its ends close, and the acceptable closing error may be large enough to conceal unacceptable gross errors in distances or angles of individual legs which only show up in the detail survey.
(v) Unless EDM is used the ground has to be cleared and be reasonably level for laying the tape and measuring the distances, and measurement is very tedious.
(vi) Unless the legs are long or the accuracy requirement low, very accurate centring of the instrument and target is required to maintain good bearings; any error in one angle is carried forward as a bearing error for the rest of the traverse.

In general traversing should only be used when triangulation is impossible, and if the area to be surveyed is only partially covered by forest then it should only be used in the forested sectors. As in triangulation the first step is to plan the network in outline on a small-scale map or photographic mosaic, having decided on the maximum spacing allowed between adjacent traverses and individual points for use by the detail surveyor and to include as many cross checks between traverses as possible. It is particularly important to include these cross checks (even at the expense of extra clearing) and not to run long many-legged traverses that approach each other away from their ends without any connection between them; the detail surveyor working later in the area between them may well find unacceptable errors between adjacent points of the two traverses. Having made this preliminary plan (probably by compass and wheel or pacing for distances) the surveyor should follow each of the routes making sure that he can establish legs of the required length; he should remember that the longer these are the fewer angle observations he will have to make, the less trouble he will have with centring, and the less computation there will be, although of course the total amount of taping will be the same. If the routes can be followed by car then this may precede the detailed location and marking of each station, but if not then these had better be done at the same time. Where a car is not feasible a bicycle can be very useful along forest paths, but good thick balloon tyres (and a puncture repair outfit) are necessary. The traverse station marks should be sited well away from any road likely to be improved, and far enough from a river or gulley bank to escape erosion. Traverse stations are more liable to damage than triangulation stations because they are more accessible, so it may be necessary to bury them completely.

Having established the ground marks and cleared the lines between them and either practised himself or trained an assistant in angle and distance measurement, the scientist can now proceed with the observations. Except for the requirement for more accurate centring necessitated by the shorter distances, angular measurement with a theodolite is the same as in triangulation. However, it is simpler because, except where two traverses join, only one angle will have to be measured and the distances are shorter so that elaborate beacons are unnecessary although the targets must be more precise. If no previous framework exists the first pointing will be to establish a starting azimuth, either to the sun† or on a bearing established by compass or some other means. Where a framework already exists the traverse should start from one of its stations, or from a triangle connected to it, and take an initial bearing from another. If this is a second station of an existing traverse the distance between the two stations should be measured and compared with their co-ordinates as a check on possible movement or misidentification. Alternatively, if two stations can be seen from the starting point the angle between them should be measured as a check that all three are correctly identified and have not moved. The same procedure should be followed at the end of the traverse when closing on another station of an already established traverse or triangulation. At the risk of repetition the need for accurate centring of target and

†The solar azimuth can be obtained from tables of the time, date, latitude, and longitude of the observation.

instrument in traversing with a theodolite must again be emphasized since it is the most likely source of error in the angles. If accuracies of 1 part in 10 000 are being aimed at, a centring error must not exceed 5 mm at either end of a 100 m leg, and if shorter legs are unavoidable

Fig. 6.3 The author plumbing a theodolite over a mark.

then the centring must be to the nearest millimetre. This can be achieved by using very small marks (such as a pencil cross or nail in a peg inserted in the hole of the ground mark) and by plumbing the theodolite very accurately over this. Fortunately most short-legged traversing occurs in forest areas and these are sheltered from the wind. If the precise mark itself at the next station can be observed, there is no problem provided that the theodolite is accurately levelled, but if not then it will be necessary to suspend a plumb bob over the mark from a tripod (from a plane table or a stand compass tripod or a tripod made from three sticks) and observe to its string using a piece of

138 MORE ADVANCED SURVEYS – PROCEDURES

coloured tape on the string and perhaps a light background to help visibility. Time spent on this for one or two legs of a short traverse will be well worthwhile in ensuring that centring errors at least do not cause the traverse to misclose. Vertical angles are required if the heights of the stations are required and also if distances between them are short enough for single tape lengths or if EDM is used. Even where the slopes of individual bays of the traverse leg are measured as part of the distance measurement, measuring the vertical angles (and hence the slope of the whole leg) at each traverse station provides a means of checking these by calculating the difference in height of the two ends of the leg.

Height measurement

The principles of heighting by three simple techniques, spirit levelling, measuring the distance and the vertical angle, and using one or two barometers, were described in Chapter 3. Of these, using the simple instruments assumed to be available at that stage, only the first has any claims to real accuracy, and no fundamental changes are required in the technique to achieve accuracies of the order of 1 cm times the square root of the distance levelled in kilometres. In general this sort of accuracy, even in flat areas, is adequate for most practical purposes, and it is only when establishing primary or secondary networks of levels for a whole country or for a large engineering project involving tunnels or canals that greater accuracy is required. Even then the techniques are essentially the same, but more powerful and sensitive instruments and special metal staves are used and more stringent precautions are taken to prevent the accumulation of systematic errors. The actual observations are very tedious, and experience has shown that for extensive networks the best answer is to employ a team of technician levellers supervised all the time by one intelligent professional or graduate (see pp. 52-3). Barometric heighting is in a totally different category since very little can be done to improve its accuracy beyond about 1 m for each height, and for more extensive projects the same basic techniques are used with more expensive instruments and more of them, and with more refined methods of observation and of computing and adjusting the results.

Thus the only real improvement in heighting accuracy which is within the range of a scientist comes from the use of the theodolite, and the main effect of this is to make possible heighting accuracies approaching those of spirit levelling in conditions where the ruggedness of the terrain makes levelling either impossible or very uneconomic. The techniques of heighting by theodolite fall into two quite separate categories which correspond roughly to the two separate techniques of establishing horizontal positions by triangulation and by traverse. The first category includes what is generally called 'trig' heighting, i.e. heighting by vertical angles observed from one (or preferably both) ends of a relatively long line the length of which is determined separately, usually as part of fixing their positions by triangulation or EDM. Since the distance is accurate the height difference can be derived accurately whatever the slope of the line. The second category comprises height traversing either in what is called tacheometry, when distances are measured by observing a graduated staff, or by the use of a wheel or chain in cases where the slopes are fairly small and accurate distances are therefore not required for determining accurate height differences.

Before describing these techniques in detail four important points about heighting measurement should be made. The first is that in nearly all engineering or agricultural design (and especially where water is involved) height is a much more sensitive dimension than length or breadth. As we have already seen in Chapter 3, field scientists can expect to be able to determine relative heights to about 1 part in 100 000 of the distance between two points of a framework by spirit levelling, while to establish the distance or bearing itself to 1 part in 10 000 is regarded as advanced surveying and is about as far as a non-surveyor may be expected to go. Even though higher accuracy in distance is relatively easy to obtain with expensive EDM equipment, it is the greater speed of this equipment rather than its accuracy which is the main advantage. Again, it is quite common to see project maps at scales around 1/20 000 where the plottable horizontal error is about 10 m with spot heights shown to the nearest metre or less and contours at intervals of 5 m or less. The second point is that, whereas in planimetric or horizontal angle observations it pays to use the longest possible sights, in deriving height differences the reverse is true. This is because, while horizontal or sideways *refraction* is generally very small (unless one is unwise enough to observe a *grazing* ray passing near a rock or building), vertical refraction is uncertain and can be very large, especially in rays passing close to bare ground in the early morning. Although its effects can be eliminated to some extent by observing from both ends of a line, the errors are still appreciable in accurate work and increase rapidly — almost as the square of the length — with longer lines. Since, as we have already seen, accidental errors accumulate as the square root of the number of observations, it pays to measure relative heights over long distances not directly but by breaking these down into a larger number of short lengths, and this process reaches its ultimate in precise levelling. For less accurate heighting than this the lengths of the sightings may be increased, but they will still be much shorter than the sides of a triangulation, and the procedure is more akin to traversing than triangulation and will be more accurate than deriving height differences from the vertical angles measured at each end of a long triangulation side. The third point is that unless a ray is fairly steep, say over $5°$, the effect of an error in distance on the height differences obtained by vertical angles is relative small, and so in height traversing, especially over terrain which is not too rugged, the horizontal distances can be measured fairly roughly — which means cheaply and quickly — by such means as observing the stadia hairs on a levelling staff, or with a chain, or even using a measuring wheel. For example, over a 100 m distance a vertical angle of $1°$ will give a 2 m difference in height, and an error of 1 m in the distance (1 part in 100) will cause a corresponding error of 1/100 in the height difference which is only 2 cm. The final point is that in height measurement good permanent marks are just as, or even more, important than in planimetric work, especially where there is a likelihood that the results of the survey may be used for the design and setting out of water channels. If enough well-defined man-made objects (hard detail) are shown on a map it is possible to locate oneself and set out a design to plottable accuracy using a compass or tape, but it is only possible to obtain an accurate height on a contoured map to about half the contoured interval, and something much more precise and certain than this is likely to be required for setting out any structure where water is involved. The map is used for the design, but the framework is essential for putting the design into practice on the ground.

The use of the theodolite for measuring horizontal angles was described in Chapter 5. The technique for measuring vertical angles in *trig heighting* is very much the same, except that it is not possible to repeat the observations on different parts of the circle and that in this case the target is intersected by the horizontal wire a little to one side of the vertical one. In general it is a mistake to combine horizontal and vertical observations; both are likely to suffer, and any delay in completing a round of horizontal angles may allow time for a movement in the instrument which will cause unnecessary errors to occur. In vertical angles it is of course necessary to centralize the fine bubble attached to the vertical micrometers or verniers using the slow-motion screw; in old instruments this may have to be done from the side or it may be possible to be done from the eye end by means of a mirror over the bubble tube or by using a prism which splits the bubble and brings the two ends into coincidence at the eye end when the bubble is central in the tube. In the latest instruments (and this includes levels also) there may be a pendulum or other device which provides automatic levelling. With all except this type of instrument the procedure must be to sight the telescope, then adjust the bubble, and then finally to check both sighting and bubble position with 'hands off', i.e. without touching the instrument. With an automatic level the observer must check that the instrument is level enough to allow this to operate. It should be noted that while observations must be taken on both faces to avoid both minor and gross errors, and while they can be repeated more than once to obtain better accuracy, there is no way in which they can be taken (as is possible in horizontal angles) on different parts of the vertical circle. This is because refraction limits the accuracy of vertical angles to such an extent that the extra cost and weight of such a device is not justified; it may also be noticed that in many theodolites, to save weight and for the same reason, the vertical circle is actually smaller than the horizontal circle. Since the actual pointing of the telescope can be as accurate as in horizontal angles, the surveyor must ensure that he has equally precise and unambiguous targets; this is best ensured in close-range work by fixing a short cross bar to the ranging rods, and in longer rays by using a blackened tin or other object on some part of the beacon with a well-marked horizontal line or edge to which observations can be taken. Observing to one edge of a ranging rod stripe has obvious dangers since the same stripe may not always be used.

In observing accurate trig heights over long distances the main cause of error, as has already been mentioned, is the uncertain vertical refraction. This can be minimized in two ways. The first is to make observations at a time when its value is least, even at the expense of some sharpness of the image, which is in the middle of the day when the atmosphere has been well stirred up by the sun. The second way is to observe the angles from both ends of the line and if possible actually at the same time of the day or even simultaneously, although the latter involves two observers and instruments and communication by radio and will normally not be practicable except in a major survey operation*. Although the early morning is the best time for horizontal observations because of the steady sharp images that are seen, it is the worst for long-range vertical angle observations especially where the ray passes fairly close to ground bare of vegetation and under clear skies. This can be confirmed by watching the movement of the image of a distant target in such conditions in a theodolite for a minute or two and seeing how rapidly its elevation can change by several minutes of arc as the rising sun warms up the cold layers of air which have accumulated above the bare ground which has been radiating out heat all through the night. A fuzzy vibrating image observed at midday through shimmer may seem less accurate but it does maintain a steady average altitude and if it is a light or helio it can be observed with an accuracy of a few seconds of arc. Thus ideally vertical angles not only have to be observed separately from horizontal ones but even at a different time of day, especially in the tropics, and the surveyor must use his judgement on how far he can afford the time to do this, the relevant factors being how long it has taken him to reach the point, the length of the rays being observed, and their importance in his heighting framework.

A brief reference was made in the discussion of horizontal angle observations to satellite stations, i.e. observing from a point which is not the same as that occupied by the target or beacon previously observed to; this should be avoided as much as possible. In vertical angle work the same difficulty is more common in that it is often impracticable (unless specially designed tripods and targets are used) to have the target height above the permanent mark or peg exactly equal to the instrument height, and both must therefore be measured and recorded and allowed for in computing the difference of height between them. It is almost axiomatic in survey work that the easier and more trivial a measurement the more likely it is to be forgotten or made wrongly, and the observer must make a conscious effort to record these heights correctly and clearly (preferably with a sketch in the field book) and not leave it to an assistant or he will find that his care in observing has all been wasted.

If all these precautions are followed trig heights of the order of 1 m can be obtained for long rays around 10 km even with a 20 second

*In the tropical dry season the weather is usually very much the same day after day, and so observations at two ends of a line made at the same time on different days may give quite good results.

vernier theodolite provided that each angle is observed at least twice on each face, while with a better instrument the accuracy can be brought to 0.5 m. In smaller frameworks with rays around 1 km long the accuracy should be of the order of 10 cm or better. Clearly it pays to connect a few of the trig heights of any extensive framework to more accurate levelled bench marks round the perimeter, and for this purpose direct theodolite observations to and from a hilltop triangulation station to the nearest bench mark − or a peg on the road near it − is certainly much quicker and probably as accurate as levelling up a steep slope to the point, provided of course that the distance is known with sufficient accuracy. In general trig heighting is inferior in accuracy to any form of levelling or height traversing, and this ties in with the requirements since accurate heights are normally required in the valleys for development purposes, while the heights of hill and mountain tops are seldom required with the same accuracy. The main use of trig heighting is in controlling general contoured surveys made either by plane table or from aerial photographs; its chief advantage is its speed, especially if a horizontal framework is already being observed, since the extra work and time involved is then relatively small.

In tacheometry or height traversing the general procedure is similar to that of levelling (and is also carried out in the tropics in the early morning or late afternoon) in that the instrument and the staff (or a special height target on a rod) leapfrog, with sights of roughly equal length taken backwards and forwards from each instrument set-up. The differences from levelling are that the line of sight is not level but sloping and this slope is measured, and that the distance is also measured so that the height difference can be calculated by trigonometry. The distance in tacheometry is derived from the length of staff intercepted by the two stadia hairs in the theodolite telescope which subtend an angle of $1/100$ (about $0.5°$) so that when the sight is horizontal the horizontal distance is exactly 100 times the difference between the readings of the stadia hairs on the staff. In a sloping ray, with the staff held vertical (and this is more reliable than trying to have it held at right angles to the line of sight), this relationship will no longer be true; but the small difference is taken care of by computation or by the tables which are generally used for deriving distances and height differences from stadia hair intercepts and vertical angles. If the positions of the instrument and the staff are required for detail survey as in minor traverse, or for distance and bearing as in contouring or profiling, then clearly the horizontal angles will also be required, but if the objective is simply to obtain by a series of shorter sights the accurate height difference between two framework points, then this is not necessary. The traverse starts exactly as in levelling with the staff or height target set up on a benchmark and the theodolite set up as far as is practicable along the route being followed. This limit will be defined either by the longest sight possible before turning a corner, or by the maximum distance at which the staff can be read, or if using a target the maximum distance at which this can be intersected with sufficient accuracy. A 14 ft (4 m) staff can be observed and read in good conditions up to 200 or 300 m, but a target on a stick can be intersected up to about 500 or 600 m so the main advantage of height traversing as compared with tacheometric levelling lies in the smaller number of set-ups; however, in general it requires more assistants because in addition to someone to hold the target the distance has to be measured by chain or wheel, which requires at least one more assistant.

Detail survey with a theodolite

Methods of detail survey using simple instruments were outlined in Chapters 2 and 3 with some indication of where and how each could best be used. This section will describe the techniques of using a theodolite as a distance-measuring instrument as well as for angles, which enables us to fix a large number of points by bearing and distance from one set-up or from two close together. Using a theodolite for detail survey, with all that it involves in weight and the time required for setting it up at each station and in computing the results, is only justified if it saves a good deal of walking, because a large number of detail points can be fixed from one set-up or pair of them, usually by bearing and distance. The secret is to use its extra accuracy in angular measurement to derive distance by optical means instead of using a tape or chain which will be much slower over any but the shortest distances. Compared with the techniques of intersection by plane table or compass from two widely separated stations, this technique has the main advantage of being able to identify each detail point fixed from only one point of view and so avoiding the difficulty (especially in small-scale plane-table surveys) of having to write or draw elaborate descriptions of each point when it is first observed so as to be sure of identifying it correctly from a quite different point of view, and some time later, at a second station. The main disadvantage of using a theodolite for detail survey is that it involves some calculation for fixing and heighting each point and the results cannot therefore usually be plotted in the field. Three reasons for using it in this way are the following: first if the weather is so bad or harsh that time or endurance available in the field is so limited that it must all be used for observation; secondly, if assistance to carry instruments is limited so that only one instrument can be carried to the top of the hills; thirdly, one or more of the hill tops gives a view over so wide an area that a large amount of detail can be fixed directly from them (this is particularly true if the number of such vantage points is limited). As we shall see below, a particularly powerful technique is possible when one or more hills overlook a sea-shore, lake, or marsh where all the detail to be surveyed is on the same horizontal plane.

The use of the theodolite in this way can be divided into two categories: first it can be used as a range finder in medium- or small-scale surveys where it is not possible to employ an assistant to occupy the position of each detail point with a staff because he would be too far away and out of touch and would take too long to move from point to point; secondly, it can be used as a tacheometer in large-scale surveys where distances are short enough to be able to tell an assistant where to go, and where the time spent by him in moving from point to point is relatively short, i.e. of the order of a minute or two. In the first case the theodolite is set up successively at the two ends of a short base and the remote points are intersected from these with great care to derive the small remote distance angle accurately; in the second case the built-in vertical angle subtended by the stadia hairs (of 1/100) in the theodolite diaphragm is used to deduce the distance from the section of a graduated staff held at the distant point and intercepted by these two hairs in the field of view as already described on p. 45. A third alternative, involving very expensive instruments, is to measure and record distances electronically, but a field scientist is unlikely to have access to such a 'total station' instrument.

Short-base fixing will normally be done from the tops of hills overlooking the survey area, and whether it is done at the same time as the observations for the framework or on a later visit will depend mainly on the scale of the map and the time required to reach and climb each of these hills; if this is not too great then the preliminary acquaintance with the area obtained by moving over it during the framework observations and seeing it from above may help in choosing the right detail points on a subsequent visit. However, on very small

146 MORE ADVANCED SURVEYS – PROCEDURES

scale or exploratory surveys or with high and difficult framework stations there may only be time for one visit, and clearly in this case all the work possible must be done then. In each case the framework point at the top of the hill should be used for one end of the base and another point of the framework should be used to obtain an initial bearing (Fig. 6.4). The base should be set off from it in such

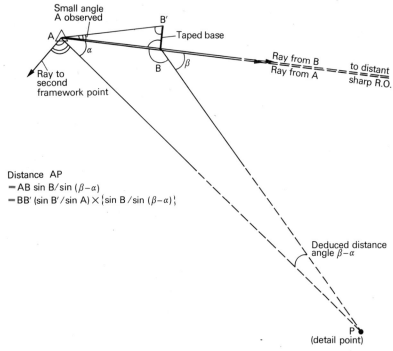

Fig. 6.4 Short-base intersections.

a direction that the majority of the detail to be mapped lies well to one side or other of the base and not beyond either end where the distance angle at it would be too small. If there is a good all-round view then three bases should be set out to form a triangle and all the detail visible can be intersected from one or other pairs of these. If the hill top is flat the length of one base can be measured directly by tape, but if it is not and subsidiary hill tops have to be used, then the length of a base can be measured by using a *subtense base*, i.e. setting up an even shorter taped base roughly at right angles to one end of it and measuring accurately the small angle (say 5° or 10°) this subtends at the other end. Alternatively, a specially made *subtense bar*, of known length between two targets, can be used but it is an additional piece of equipment and is not really required. The length of the main base should be such that the most distant and oblique detail points subtend angles of at least two degrees using a one second theodolite, or 5°-10° using a less accurate instrument.

Having set out each base and measured its length, the next step is to decide on the detail points to be fixed, list them, and mark them on a sketch panorama or a map or on a round of Polaroid-type photographs. The points must be sharp enough for accurate intersection. Rounds of calibrated photographs can also be taken for enlargement and used later in interpolating features between the detail points when these have been plotted on the map (details are given on p. 71). The theodolite is set up accurately over one end of the base and its horizontal circle is set to near zero when pointing at the other end. Over such a short distance very accurate centring would seem to be essential if the remote angle (which is derived by subtracting the angles at the two ends of the base from 180°) is to be obtained with sufficient accuracy, but in fact this difficulty can be overcome as follows (see Fig. 6.5). A sharp distant unambiguous object is selected beyond one end of the base such that it can be seen from both ends and is on the line of the base, and angles to the detail points are measured from this as a reference object and not from the other end of the base. If this is not possible then both ends of the base must be marked with very fine marks (such as a nail in a peg as in short-leg traversing (see p. 137) and great care must be taken to ensure accurate centring of the theodolite at both ends of the base. If, for example, the deduced angle at a detail point is about 2° and its distance is 10 cm on the map, then this distance must be measured with an accuracy of 1 part in 200 to achieve a plotting accuracy of 0.5 mm. One two-hundredth of 2° is approximately 0.5′ and over a base length of 100 m this is 1 part in 6000 or about 2 cm, so centring should be accurate to at least 1 cm at each end which will require the use of a screen for the plumb bob in windy conditions or of an optical plumbing device fitted to modern theodolites. Angles should be measured to about ten seconds of arc which means that the points must be observed to on both faces even with a one second theodolite. If this is not available the base must be lengthened so that the distance angles are of the order of 5° or more; clearly an angle of even 10° will still cause no difficulty with identification of the points from the two ends of the base.

148 MORE ADVANCED SURVEYS – PROCEDURES

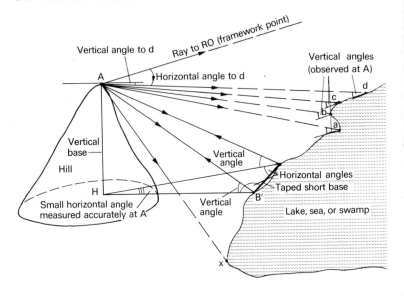

Fig. 6.5 Use of the vertical base.

Heights can be obtained for all the points fixed by observing the vertical angles, and even if their heights are not required it may be worth measuring these vertical angles accurately because the two heights obtained for each of the detail points will then be a check on the observations, the identification, and the calculations and plotting of the positions – in all of which mistakes can occur with a large number of points. In any case, the slope of the base, if it was measured with a tape, must of course be observed to about the nearest minute of arc so that the true horizontal length can be obtained; however, if subtense measurement is used then this is only necessary if heights are being observed. Since all other distances are derived from horizontal angles measured with the theodolite, it is the true horizontal distances to these which are obtained. However, if the distances to the detail points had been measured directly by range finder, then of course they would be the slope ranges and corrections would have to be applied before plotting them on the map. In very rugged country a vertical base (with one observing station a long way above the other) could in theory be used, but in general this is unlikely to be practicable. However, one form of vertical base is very useful when surveying detail in a very flat area such as a sea-shore, a marsh, or a lake overlooked by a high hill or mountain. This can be a very powerful technique and save hours or days of tedious traversing, plane tabling, or compass survey if one is lucky enough to have one or more viewpoints sufficiently high and near to derive distances from them to points on the level surface by simply measuring the vertical angles from each viewpoint to those visible from it. The exact height of the theodolite at each station must be obtained above the level surface, and the easiest way to do this is to set out on this a short base marked by two poles and then to observe the angle subtended by these at the hill-top stations together with the vertical angles to and from them. The horizontal angle at each end of the measured base below should also be observed and this should be as near a right angle as possible (Fig. 6.5).

A series of single pointings observing both horizontal and vertical angles is then taken from the hill top to each of the detail points required at the edge of the water surface or in the marsh, and if the vertical angles are large enough these can be on one face only which saves a great deal of trouble in recording where the points are. However, as with other detail surveyed by theodolite, it pays to draw a sketch outline of the shore or other flat features or boundaries first and to mark on this the approximate positions of the points. These should be lettered or numbered as in the angle book so that the features required can be correctly interpolated between them. For a given height above the level surface accuracy falls off as the square of the distance, as shown by the following example. A hill 100 m above a lake or marsh should have its height measured to about 20 cm or 1 part in 500. The height required is that of the actual theodolite axis so that if an existing ground mark height is used the extra height to the instrument's horizontal axis must be added. Observing to a detail point at a distance of 1 km would give a vertical angle of 1 in 10 or about $6°$ which is $360'$, so this should be measured with an accuracy to better than a minute; the computed distance would then be accurate to about 2.5 m which would be less than 0.5 mm at a plotting scale of 1/10 000. However, a point at a distance of 2 km would give only a $3°$ ($180'$) vertical angle; with a comparable accuracy of angle measurement the proportional error in the distance would be 1/180, and over 2 km this would be over 10 m or four times as much. In very small scale work, with heights of several hundred metres and distances of 5 km or more, allowance has to be made for the earth's curvature and for refraction.

150 MORE ADVANCED SURVEYS – PROCEDURES

Clearly this technique of short or vertical base intersection requires both a feeling by the surveyor for the limits of error in the various processes and a fair amount of computation, so it is more suitable for those with some mathematical ability and intuition. However, modern pocket calculators which give trigonometrical functions directly on the keyboard make this kind of computation very much easier and less liable to mistakes than using either logarithms or tables of natural trig functions and a hand calculator, and this has materially altered the balance between graphical and instrumental/computer techniques in the last few years. When observing to the sea-shore the effect of the tide must not be forgotten, and either all observations should be taken to a common high-tide mark or allowance must be made for the height of the tide if observations are taken to the water's edge.

Tacheometry

In large-scale work tacheometry is employed using a staff, and the distance is derived directly by a single shot to each detail point (see Fig. 6.6(b). As mentioned on p. 145 the theodolite diaphragm contains, in addition to the main vertical and horizontal wires (in modern instruments lines engraved on glass) two short stadia lines placed symmetrically on the vertical wire so that they subtend an angle of 1 part in 100 at

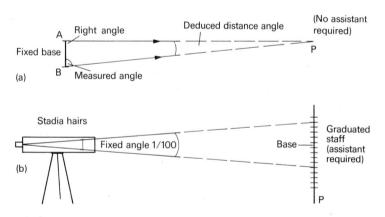

Fig. 6.6 Optical distance measurement by range finder and tacheometer (theodolite or level). (a) Range finder: fixed base; small variable distance angle deduced by measuring the angle at B. (b) Tacheometry: fixed angle; variable distance observed on staff.

the eyepiece. The intercept on a graduated vertical staff gives the distance, with an error which is 100 times the reading errors on the staff. Thus an intercept of 1 m, read to 0.5 mm, gives a distance of 100 m with an error around 5 or 10 cm. At longer distances, with similar reading accuracy, the error would be the same, but in fact of course the reading accuracy decreases and so the error at 200 m which is near the practical limit, will be 20 or 30 cm. Similar stadia hairs are supplied on level and telescopic alidade diaphragms so that these can also be used for measuring distance in this way at comparatively large scales. Except when using a level it is of course the slope distance which is measured, and so this must be corrected to the horizontal. Moreover, unless the staff is held at right angles to the line of sight or the slope is very small, there must be a correction for its inclination to this, and from simple trigonometry it can be seen that the true horizontal distance L and the vertical distance dH (the height difference) between the theodolite and the staff are given by

$$L = (R_1 - R_2) \times 100 \cos^2 V$$
$$dH = (R_1 - R_2) \times 100 \sin V \cos V$$
$$= (R_1 - R_2) \times 50 \sin 2V$$

where R_1 and R_2 are the staff readings and V is the slope angle. Values can be read from tables or from a hand calculator. The vertical angle must be measured to a point of similar height on the staff, and if the height of the instrument station is also required at each set-up of the theodolite or level the staff is stood alongside this and the height of the telescope marked on it by a piece of coloured Scotch tape to which the main cross wires of the telescope are always pointed. The exact reading of this on the staff should always be recorded at each observation because this acts as a check on the other two readings. It should be almost exactly midway between them, except that on steep slopes the intercept further from the instrument (the upper looking upwards and the lower looking downwards) will be slightly longer than the other.

Range finders

Where no assistant is available distances can be observed by range finder (see Fig. 6.6(a)), although few professional surveyors think this worth its weight. Since it has a very short base (usually a little less than a metre) and moving prisms, it is very easily put out of adjustment; the error increases as the square of the distance and this is indicated by the rapid diminution of the divisions on the scale at longer ranges. The instrument must be checked each time it is taken out of its case, either by measuring the range to another fixed point of known distance or by observing to two pins fixed in a base frame of equal length provided with the instrument. Another disadvantage is the difficulty of obtaining precise coincidence of the two images unless they are of a sharp hard vertical feature such as the corner of a building, or a telegraph pole, or a straight tree trunk. The main use of this instrument is in reconnaissance mapping in rugged country when bearings or angles are taken by compass or plane table.

An outline of computing procedures

Triangulation

The computation of a triangulation framework observed by theodolite should be in three stages: first checking and if necessary adjusting the angles, secondly computing and if necessary adjusting the side lengths, and finally computing the co-ordinates, although this may also be in two stages – provisional, followed by final results. The first step is to make sure that the angles are correct. A final trig diagram should be compiled from the reconnaissance and observing plans (Fig. 6.1) and checked against the field books; if the reconnaissance plan was very approximate the final trig diagram should be produced by plotting from the actual angles observed using a protractor. Rays are shown solid if observed from both ends, and are pecked at the target end if not. The measured lengths are shown by a double line (or by ticks across the line), and any major figures to be treated separately from the rest are shown by heavier single lines. Existing and newly occupied stations are shown by triangles, and intersected (unoccupied) points are shown by circles. The mean of each angle taken from the field books should then be checked by closing all the figures (triangles or quadrilaterals), and the angles tabulated both as observed and as adjusted. In triangles one-third of the misclosure is normally taken from or added to each angle except where one of them is under $30°$ and has been observed on more zeros than the others when the adjustment to it should be (say) half that of the adjustment to the other two. In a quadrilateral with both diagonals observed there are still two further adjustments, even if all the observed triangles are closed, because with the second diagonal we have one redundant observation and the whole quadrilateral must be made self-consistent. First the deduced angles where the diagonals intersect must agree, and secondly the side lengths if computed from different triangles from the common side must also agree. It is inadvisable to include a figure more complicated than a quadrilateral. Closing errors of triangles should be about 20″ for work done with a 10″ reading micrometer theodolite, and up to 1′ with a vernier instrument reading to 20″. Errors larger than this indicate either a centring error of some kind, or possibly too much rejection of apparently anomalous readings of each angle (see p. 78). In very large triangles, where the curvature of the earth becomes appreciable, the angles add up to 180° 00′ 00″ plus *spherical excess* which increases with the area of the triangle. It is normally not appreciable.

Once all the figures have been closed and the sea-level length of the bases (or the lengths computed from given co-ordinates of any existing framework sides in the area) have been obtained, computation of the side lengths can proceed through the framework starting from the longest and most reliable base or side and using the simple sine formula $a/\sin A = b/\sin B$. It should proceed from the first base to the second by the best route using a sensible combination of the fewest and largest figures but also those with roughly equal angles (for triangles near 60° and for quadrilaterals near 90° and 45°) since these will be the strongest figures for carrying forward the lengths. This process will eventually give a computed value for the second base (or triangle side of a superior framework) which will differ from the measured length of the base reduced to sea level (or the one computed from its end co-ordinates). Provided that this is within the sort of accuracy expected, i.e. about 1/10 000 with a micrometer theodolite and 1/5000 or so with a vernier one, no adjustment is necessary and the rest of the side lengths can be computed directly from the side lengths already obtained. Only if the framework is extensive and the scientist feels able to do it need a length adjustment be made by distributing the progressive scale error equally between the two given lengths.

In deciding whether to do a more elaborate adjustment a simple rule of thumb can be used. Would the apparent scale error between the bases produce a plottable error in any part of the map? If the first accuracy (1/10 000) has been achieved, the complete map would have to be about 5 m long to produce a plottable error of 0.5 mm anywhere on it, and in the second case it would have to be over 2 m long. However, not adjusting the lengths before proceeding with the detail survey does not mean that it should never be done. If the framework only covered one map sheet, was observed with a vernier theodolite, and was marked only by pegs, then a second adjustment is not worthwhile, but if it was carried out with a micrometer theodolite to the sort of accuracy quoted above and if the stations were permanently marked, then it is. In this case it is best to hand over all the *original observations or photostat copies* (*not* repeat *not* hand copies or summaries which introduce errors) to a professional land surveyor or computer for him to carry out a proper adjustment, probably using an electronic computer into which he can feed all the equations which would give the best possible result from a larger number of observations, some of which are redundant. The provisional co-ordinates computed by the surveyor from unadjusted observations will be good enough for field plotting and possibly for the final map, but it would be a pity to spoil an excellent ship for a ha'porth of tar; if the framework has been well observed and marked it is capable of use at very much larger scales by the engineers and cadastral surveyors who will be working in the area if the agricultural or other developments envisaged in carrying out the original survey do in fact materialize. It would be a pity if they had to repeat all the work because the relatively small cost and effort of adjusting and computing it properly had not been done and the original observations had not been preserved. This two-stage approach is followed by professional land surveying organizations (such as the UK Directorate of Overseas Surveys) since in their case, particularly when producing 1/50 000 scale maps, errors of several metres are acceptable in provisional co-ordinates for the mapping, while for the country's national framework which is required for long-term use possible errors at any one point must be reduced to centimetres by proper careful independent checking of all the observations and by adjusting them mathematically.

Having checked the angles and side lengths in this way and made sure that there are no gross errors, the scientist can now proceed to compute co-ordinates for plotting his points on the field sheets preparatory to carrying out the detail survey. Of course it is possible to plot the points directly from the angles and one or two side lengths, and if the survey is confined to one map sheet or closely controlled by an existing framework, this may be adequate and will provide a check on gross errors. A large protractor (at least 20 cm in diameter) should be used and a very hard pencil with a sharp point. The disadvantage of this method is that it depends on a successive build-up of triangles, and so the absolute positions of points between the two bases or any check points may have perceptible errors created in the plotting alone; also the cuts at each point will normally not be at right angles and this produces a 'distance' error. Provided that an accurate grid has been drawn on the field sheets (and this can be done by a professional draughtsman before going overseas or in the field), then each point plotted from co-ordinates is independent of the others so that the drawing errors are not cumulative; also each point is plotted by two lines at right angles giving the best possible result. Moreover, with modern programmable computers costing less than £100 and providing trigonometrical functions (sine, cosine, and tangent) directly for automatic multiplication with the distances, most of the labour and sources of error caused by computing with logarithmic or natural functions taken from tables have now disappeared. Advanced textbooks, such as Olliver and Clendinning Vol. I, should be consulted.

For an extensive project, therefore, the triangulation must be plotted on a rectangular grid, and this at once raises the question: what grid? There are two possibilities. If the survey is based on an existing framework then the co-ordinates of the points used will already be available on the national or some other extensive grid which is likely to be on some sort of a *projection*. This is necessary because the curved surface of the earth cannot be represented on a flat sheet of paper without some distortion either of dimensions (scale) or of shape. A *conformal* projection (like Mercator's) preserves local shape but distorts scale especially near the poles (which never appear on it) which is why it makes Greenland as large as Africa and why it can scarcely show Antarctica at all. Meridians and parallels are everywhere at right angles on it as they are on the earth, and so, although the scale will differ from one point to another, over any small area it is equal in all directions and thus local shapes and angles are correct, i.e. the projection is conformal. At the other extreme we have *equal-area* projections in which scale is preserved but local shapes are not; these are used in atlases and small-scale maps for such purposes as comparing population or other densities. Surveyors use conformal projections, and the most common is a form of Mercator called the *Transverse Mercator* of which a special variant is the *Universal Transverse Mercator* which was worked out in the USA and is being adopted increasingly everywhere except in very high latitudes. In this projection angles at each point are true but the scale varies from east to west, being adjusted to be correct at two meridians each one-third of the distance from the edges of each zone which covers 6° of longitude. In this way the scale error never exceeds ±1 part in 2000 anywhere which is not enough to bother the detail surveyor and is often exceeded by shrinkage or expansion of the paper or other map material. None of this need really bother the field scientist computing his work from existing trig sites on a national grid (which is another advantage of trig over traverse), but he must realize that the lengths of his framework will differ from any lengths measured on the ground (even if these are also reduced to sea level) by small amounts of the order of 0.5 m per kilometre. Angles in large triangles can also differ slightly from their true values, but only by a few seconds and they can be safely taken at their adjusted values. Unless the area is very large the scale error can be taken as uniform throughout.

The second possibility where no existing framework has been used is to use a local or arbitrary plane grid approximately oriented to true north by compass or to use astronomical observations (which are beyond the scope of this book). In this case there are no scale corrections, but distances should be reduced to sea level before starting computation of co-ordinates. The starting point is given arbitrary co-ordinates. These should not be zero or the same for easting and northing, first because this may mean both positive and negative co-ordinates for the remaining points, and secondly because northings and eastings are easily mixed up if they have similar values. Thus a good starting value is 100 000 m east and 50 000 m north, assuming that the start point is somewhere near the south-west corner of the area and that this extends for less than 50 km in both directions. A sensible adjustment to these figures should be made for other circumstances so that the most south-westerly point has suitable co-ordinates. Having determined (or assumed) the initial bearing and length of the first side, values for the co-ordinates of the other end are now easily computed using its length multiplied by the sine for eastings and the cosine for northings, assuming that the bearings are all taken right handed from true north. The correct sign of the differences must be allotted before adding or subtracting them from the starting co-ordinates, and a diagram is best for this. This first computation must be very carefully checked or repeated because all depends on it, but thereafter, as each third point is computed independently by bearing and distance from the other two apices of the triangle, a check is provided by the double computation, starting from two different points and using quite different bearings and lengths to arrive at the same third (unknown) point.

A traverse network

Having checked and corrected the measured lengths, computing a traverse starts in the same way as a triangulation by checking and adjusting the angles, but no separate length check is possible and so the next step is to compute the co-ordinates through the traverse and compare those obtained for the last point on which it closes with the given co-ordinates of that point (which in a closed traverse will be the actual starting point). The angles are used to obtain the bearing of each successive leg, and the angular misclosure with the final bearing is then distributed equally through the traverse (assuming that it is within the acceptable limits). The usual criterion for the closing angular error is to determine an acceptable error for each angle and multiply it by the square root of the number of stations. Having distributed the closing error through the bearings of the legs, the co-ordinates of successive stations are then computed using these and the horizontal sea-level values of the measured lengths, if necessary corrected for local scale error of the grid. Where the taping has been in catenary the sag correction must be applied to all sections as described on p. 124. Even after using adjusted bearings the co-ordinates of the final station will normally not agree with the given ones, and the length of the closing error (computed from the square root of the sum of the squares of the closing errors in easting and northing) should be compared with the total measured length of the traverse to see whether it is within the proportional error allowed. If it is outside this the direction and magnitude should be compared with the directions of the individual legs. If it is parallel to one of them then a gross error in measured length or in computation or in copying may be suspected, particularly if the error approaches some value such as a tape length or 10 or 100 m. If the error is at right angles to the general direction of the traverse, then an error in angular measurement or in centring may be the cause, and the observations at the ends of the most likely leg or legs should be repeated. If no gross errors are found which reduce the closing error to acceptable limits the whole traverse may have to be repeated.

Whereas in a triangulation, after checking all the angles and closing the figures, the next step is to compute the lengths through from one base to the other by the shortest route and thus be sure that both angles and lengths are correct before computing any co-ordinates, in a traverse network the only independent check on the distances is obtained by computing co-ordinates of the same point by two different routes and comparing them. The first traverses to be computed should therefore be either closed circuits or those that run between points of an already established framework of a higher and more reliable order, which may be a triangulation observed and computed by the scientist himself. These 'master' traverses must be computed and, if closing satisfactorily, adjusted first. If no framework points of a higher order exist then the first step must be to compute a traverse round the area closing back on itself. If this closes satisfactorily then its co-ordinates can be adjusted provisionally, but should not be accepted until one or two traverses running diametrically across the area by the straightest possible routes have also been computed and are also seen to close into it satisfactorily.

The scientist doing this work himself will at once begin to appreciate both the general disadvantages of traversing, because of the late application of independent checks, and also the necessity for being able to trust his own observations. All the tedious precautions against gross error advocated above, such as measuring all horizontal angles at least twice with different starting values, measuring all distances twice, preferably in different units, accurate centring of stations on short legs, checking the copying of any figures in computation and so forth will now be seen to have been worthwhile if they mean that all the checks work out satisfactorily. Once a gross error, or more than one, has been made, it may be very difficult to locate it at this late stage and to know what observations have to be repeated in order to achieve satisfactory closures; it is worth a great deal to make sure they are right first time.

PART III
Surveying from air photographs

7. THE CHARACTERISTICS AND USE OF AIR PHOTOGRAPHS FOR PLANIMETRIC MAPPING

Characteristics

How they are taken

For most practical purposes air photographs are taken professionally from a special aircraft and with a specially designed aerial camera, usually with the lens pointing vertically downwards. It is almost universal now to take these with a 9 in square format and with either a 12 in (30 cm), 6 in (15 cm) or 3.5 in (9 cm) focal length lens. Photographs taken with these lenses are referred to as normal, wide, and superwide angle, and of these the second − the wide angle − is by far the most common. The focal length is usually given in inches or millimetres on the title strip of each print. For military reconnaissance much longer focus lenses are sometimes used in order to achieve a larger scale from a greater altitude but these cover specific targets.

The more common vertical air photographs are taken by the aircraft in a series of *strips* each consisting of a series of photographs with the camera set so that the sides of the photographs are parallel to the direction of the strip. The camera is operated at regular intervals so that the area covered by each photograph overlaps extensively on to the one following (normally by about 60 per cent), and successive strips are flown roughly parallel so that they overlap laterally. Thus every point in the area is covered by at least two photographs; in each photograph there is a narrow band (called the *supralap*) across the centre at right angles to the line of flight covered by the preceding and succeeding photographs, i.e. three in all, and at the edges of the strips there are small areas covered at times by four, five, or six photographs depending on whether the overlaps on each strip are in phase or not. When the films are developed and prints are made and laid down in their correct relative positions they look like the scales of a fish or the tiles on a house − which are also designed for complete coverage!

The object of this system is twofold; first it ensures that every part of the area is covered by a minimum of two photographs, which can therefore be used to view any part of it stereoscopically, i.e. in three dimensions. The second is to provide a means of linking the photographs together by means of the images of small objects or points of detail at their edges or in the corners — such as bushes, path intersections, corners of buildings, small rocks, patches of dark grass or light bare earth, huts, trees, etc. — which appear on three successive photographs of one strip and also in photographs of the adjoining strip. Using the geometry of the photographs it is then possible to build up a framework which accurately relates the photographs to each other and makes it possible to plot all the details from them in their correct relationships. For this purpose the unit from which detail is mapped is the *overlap*, i.e. the approximately rectangular area covering less than half a photograph which is overlapped by the one taken immediately before or after it.

The modern aerial camera is a very complex piece of machinery costing several thousand pounds; modern films are very stable so that the size and, more important, the shape of each negative are known to within fractions of a millimetre. In some cameras a *reseau* (grid of small crosses) is printed on the film at the time of exposure in order to check on any subsequent changes in the shape and size of the film. Modern emulsions are so fine grained that the film or prints can be viewed — or enlarged — at up to six times the original scale before the grain begins to show and the detail to blur. The film is held absolutely flat at the moment of exposure; the exposure is short enough so that the aircraft's movement during it is insufficient to cause discernible image movement, and the lenses are so well constructed that although one type may cover an angle of $120°$ across the diagonal (in a superwide angle photograph) the position of every part of the image on the photograph is the same — within a few thousandths of a millimetre (micrometres) — as if the photograph had been taken by a pinhole camera.

With normal professional flying and if the country is not too rugged, the strips will be reasonably parallel, the correct forward and lateral overlaps will be preserved, and the camera axis will have been within a couple of degrees of the vertical in all exposures. Scales can vary from about 1/80 000 taken with pressurized aircraft to 1/2000 with helicopters or slow-flying fixed-wing aircraft. Where parts of the negative are very dense or thin (owing to snow patches or shadows caused by relief or clouds) so that detail disappears in the print, a mask can be made to give different exposures to these areas; or more commonly in commercial firms a scanning spot printer can be used to obtain satisfactory prints. The spot is received by a photoelectric cell which automatically and instantaneously adjusts the strength of the light to suit the local density of the negative. The result is a rather 'flat' print in which light and dark contrast is much reduced, but any detail appearing on the negative will also appear on the print, even in very light or dark areas.

Oblique photographs

For some purposes (particularly during the last war and in the polar regions) oblique aerial photographs have been taken and some of these are still available. An oblique photograph is one taken with the camera axis designedly tilted substantially from the vertical. In a high oblique the horizon appears in the photograph; in a low oblique it does not. In general, obliques were used for one of five purposes.

(i) Low obliques taken with long focus lenses were taken in pairs by the Royal Air Force often in conjunction with wide-angle vertical photographs in order to provide simultaneous cover at larger scales for interpreting detail which could then be mapped from the verticals.

(ii) Single obliques were also taken in war time for reconnaissance purposes or to record the dropping of bombs or artillery fire etc.

(iii) In peace time a considerable number of low and high obliques have been and are still being taken of factories, archaeological sites, private houses, cities, etc. for publicity or to assist in planning.

(iv) High obliques were taken before the war by the Canadians of their northern area and by the Norwegians and Danes of Spitsbergen and Greenland for mapping purposes, because the aircraft of those days could not fly high enough to carry out vertical photography at a sufficiently small scale for the sort of exploratory mapping – at scales less than 1/100 000 or more than 2 miles to the inch – which was required.

(v) A special case of high obliques is the *tri-camera* or *trimetrogon* system in which three 6 in cameras are used: one vertical and one on each side with their axes at right angles to the line of flight and depressed 30° below the horizontal (i.e. a tilt of 60°) so that each oblique includes the horizon and both overlap onto the vertical. The main advantage of this kind of high oblique is that far fewer are required to cover a given area from a given height and so flight lines can be much more widely spaced.

This system was used in the 1940s by the U.S. Geological Survey to map several million square miles in the Middle East and northern Africa for producing 1/M scale aeronautical charts, and it has since been used by them to cover very large areas in the Antarctic. The vertical photographs in this series can of course be used like any other verticals in the narrow strips they cover, but in general, if measurements or maps are required, the obliques can only be handled successfully by cartographers or surveyors.

Clearly the main use of oblique photographs to the environmental scientist is when they are the only previous cover available for studying major changes in the vegetation or the landscape, particularly of coastlines, river channels, or dune fields all of which can change substantially in a few decades. The U.S. Geological Survey photographs of northern Africa, with their verticals covering about a quarter of the area, are particularly valuable for this purpose.

Obliques (especially high obliques) may be of more value than vertical photographs for a few purposes. Because they cover a larger area they give a more general view and sometimes make possible the understanding of the structure or layout of some extensive feature like a drainage system, or a large geological structure, or a city. If cleverly taken they may highlight some characteristic of these features, but in general this is likely to pervert the truth rather than reveal it. They have the advantage of giving a familiar viewpoint since the objects look much the same in high obliques as they do on the ground or from the top of a hill, while the vertical viewpoint is an unusual one to most people who have not used air photographs before. One might say that obliques form an easy introduction to the serious study of air photographs — which must be done with stereo-pairs of verticals. One could also call them the lazy man's approach in the same way that topographic models have to be made (of municipal projects or proposed motorways) for those members of committees or the general public who are unable to understand and use contoured maps.

Relating air photographs to the ground

Vertical air photographs are invaluable for virtually all studies of the ground and its features — whether natural or man made — because they show every object which is visible in a way that no map can ever do. However, they are not automatically related to the ground by a grid or graticule or by the names of the features on them, and unless

1. Using a parallax bar with a mirror stereoscope.

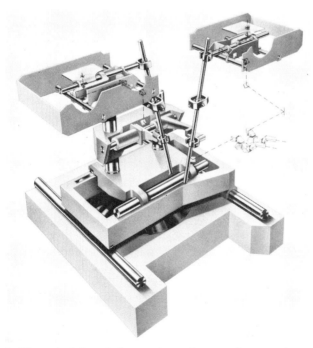

2. The principles of the mechanical type of stereo-plotter.

3. Wild Rectifying Enlarger showing how both the negative holder and the enlargement table can be tilted in equal and opposite directions relative to the lens.

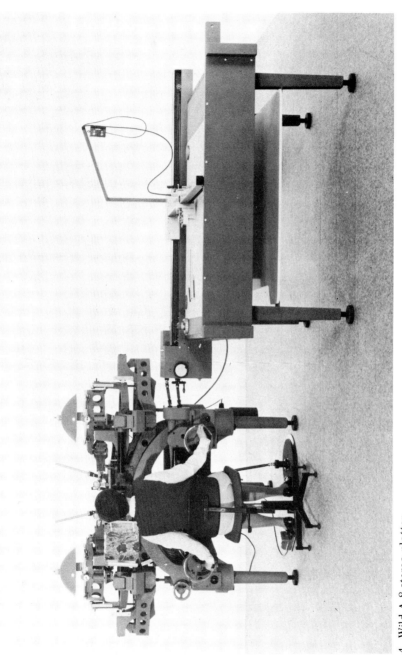

4. Wild A.8 stereo-plotter.

5. Vertical air photograph of Parliament Square London showing the radial height distortion of high buildings away from the centre of the photograph. (Courtesy Meridan Air Maps Ltd.)

this can be done they are of little value. Thus their position, orientation, scale, lack of verticality, and the moment when they were taken must all be determined at least approximately before they can be used for scientific purposes and not just as pretty pictures. In later sections more accurate techniques for locating and scaling them for compiling a map will be considered, but in this section we envisage the field scientist at an earlier stage either trying to select the right photographs to cover an area in which he is interested, or finding out what area is covered by the ones he has already. To do this he has to be able to answer the following questions about any photograph:

(i) Where is it?
(ii) Which way round is it?
(iii) What scale is it?
(iv) Is it tilted?
(v) When was it taken?

One general point should be made: most aerial photographs include along one margin a set of images of instruments showing the time, the altitude, a circular bubble, the date, the location, etc.; and many blocks of photographs are accompanied by flight diagrams showing where they were supposed to have been taken. The information given by the instrument panels and the flight diagrams should be treated with great reserve unless they are produced by a well-known professional air survey firm; amateurs (including military pilots and navigators trained for fighting or reconnaissance) often make the most astonishing mistakes about their locations in an aircraft. My own experience of this has included photographs of Zaïre said to be of the Sudan, a supposed sighting of land in the middle of Greenland which turned out to be part of the coastal area, and supposedly vertical photographs which had a tilt of over $20°$. As in all surveying operations the surveyor should be like St Thomas and believe nothing without checking it himself.

(i) Where is it?
For the ordinary non-surveyor scientist the location of an air photograph taken professionally is most easily obtained from the agency supplying the photographs, who will also normally provide a *photo index*, i.e. a small-scale map showing the layout of the photography. In some developing countries (especially those mapped by the Directorate of Overseas Surveys) the principal points or centres of the vertical

photographs are shown precisely by small numbered crosses on the maps produced from them, and the full data for the contract, sortie, and individual numbers are shown in the margin or at the sides of the map. These correspond to the individual and other numbers on the photographs themselves. Where no such data are available, the user must obtain as many consecutive photographs together as possible and compile his own index, and where the index has not been compiled by professionals he should treat it with reserve as mentioned above.

(ii) Which way round is it?

Obviously when this has been done the orientation of each photograph is also known because if it is the wrong way round the detail will not match on the overlaps. Normally the title strip on the photograph is parallel to and not at right angles to the line of flight. If the aircraft flew in opposite directions on different flight lines these title strips would be on the opposite sides of photographs of these strips, and the individual numbers would of course also run in opposite directions along them. With a single photograph or a pair the only solution is to have a map of sufficient accuracy for comparing the topographic details with what appears on the photograph(s) and then to locate, orientate, and scale them from this.

(iii) What scale is it?

This is best answered by identifying correctly features which are obviously the same on a photograph or print laydown as on a map and comparing the distances between them with the distances on the best available map. Where no maps exist, the data on the title strip must be used, allowing of course for the height above sea level of the land appearing in the photograph, and treating this with reserve until a better check is available.

(iv) Is it tilted?

Unless the horizon appears in the photographs there is no simple way of determining *tilt* (the bubble shown in the margin is *not* reliable†), and in fact for the ordinary user this is a good deal less important than the location, the orientation, and the scale. However, it is sometimes important to know for certain whether an apparently vertical photograph is badly tilted, and this can be checked if a good map of

† Even if the aircraft is steeply tilted the bubble could still be central because when making a correctly banked turn the acceleration is still perpendicular to the floor.

the area covered by the photograph is available and if the ground is also reasonably flat. If the photograph is a near vertical then the scale everywhere should be much the same; if it differs appreciably from one side to the other and the ground is not sloping then the photograph is tilted. In developed areas distortion of obviously flat rectangular features (like tennis courts or football grounds) will also be apparent. Badly tilted photographs may also show considerable differences of scale in the overlaps both along and at right angles to the flight lines. More precise methods of locating, orienting, scaling, and measuring the tilts of the photographs are required if maps, especially contoured ones, are to be made from the photographs, and an outline of these is given later (Chapter 9).

(v) When was it taken?
The date is usually given on the title strip, but these are quite often illegible and are not always very reliable in any case. Where these do not appear reference must be made to the agency responsible, and if this also fails internal evidence of the photography must be used, e.g. new roads or other developments which can be dated.

The geometry of vertical air photographs

The geometrical difference between a vertical air photograph and a map is that the map is on an *orthogonal* projection, i.e. every object on it is viewed as if by parallel lines of sight from an infinite distance, while in a photograph they are all viewed from a single point, i.e. it is a *perspective* projection. Thus, if the camera lens covers a wide angle, although objects directly under the camera at the moment of exposure are viewed vertically from above as on a map, those at the edge of the format are viewed at a considerable angle with the vertical. In a camera with a 6 in (wide-angle) lens and a 9 in wide film this angle can be over 45° in the corners, and with a superwide-angle lens (about $3\frac{1}{2}$ in (88 mm) focal length) it can be up to 60°. The effect of this is shown in Plate 5 where the clock tower of the House of Parliament appears to be leaning over considerably and Westminister Abbey in the centre does not; this is also shown diagrammatically in Fig. 7.1. There are two reasons for using a wide-angle lens: first to cover the widest possible area at the smallest possible scale from a given flying height, and secondly to provide the strongest possible stereoscopic effect. Only thus can air photographs be used for measuring heights and contouring,

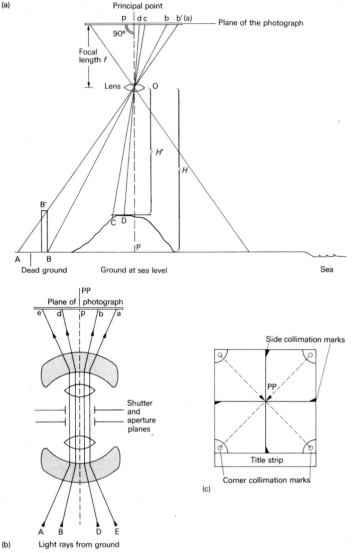

Fig. 7.1 (a) The geometry of a vertical air photograph. Note (i) the displacement of the image b′ of the top B′ of BB′ to a and so apparently to A on the ground, (ii) the change of scale between $ab/AB = bO/OB = f/H$ at sea level and $cd/CD = cO/OC = f/H'$, at the hill top, and (iii) that the scale is given by the focal length f divided by the height of the aircraft above the ground. (b) The geometry of a survey camera. (c) The elements of the photograph.

as we shall see in detail in Chapters 8 and 9. Superwide-angle photographs are used to obtain a smaller scale of photography for topographic mapping, or when aircraft limitations or, as in West Africa, dust haze make high-level photography impossible. Longer-focus lenses (12 in or 300 mm) are used for photographing towns where contours are not required and to reduce the 'lean back' of high buildings, and also in areas where low flying to obtain a sufficiently large scale with a 6 in lens may be prohibited. In the extreme case of a military requirement for large-scale 'spy' photography from a very great height, telescopic lenses with focal lengths of 1 m or more can be used. However, for nearly all practical purposes aerial survey photographs are taken with a 6 in (15 cm) lens, and this is the type which the field scientist is most likely to meet.

The requirements for a wide-angle lens, coupled with the usual criteria of any good camera for even illumination over the format, good sharp resolution everywhere (down to about 5–10 μm*), and a reasonably large stop for a short exposure, make air survey cameras very expensive, especially since there is an additional requirement that they shall be almost free of *distortion*. This means that the rays coming in even at the extreme angles from ground features which will appear in the photograph corners must, after passing through the lens assembly, emerge with their directions virtually unchanged, the differences being about 10″. As mentioned earlier the camera lens must act like a pinhole, while at the same time letting in light through an aperture which may have a diameter of up to 2 cm, for the exposure must be short and simultaneous over the whole format otherwise there will be appreciable movement of the images due to the movement of the aircraft during the exposure and the whole perspective principle will break down. This of course precludes the use of a *focal plane* shutter which takes an appreciable time to traverse the film, and the considerable size of the aperture also requires the use of a complicated rotating shutter which is much more elaborate than that of an ordinary hand camera (see Fig. 7.1(b)). The large film format, 9 in square, of the ordinary aerial survey camera also creates problems since the film must be flat everywhere to within a few micrometers, and this is achieved either by pressing it forwards against a glass plate or by sucking it back against a special *vacuum back* with holes in it. These requirements are mentioned so that the reader may be aware of some of the problems of taking good aerial

*1μm is one millionth of a metre or one thousandth of a millimetre.

survey photography. To those problems inseparable from operating an aircraft and of the weather are added those of ensuring that the camera is operating correctly all the time. Nevertheless, in what follows we shall assume that the photographs have been taken within 3° of the vertical so that each photograph is taken with the camera axis very nearly vertical, with the film flat, and with the lens free of distortion, and that the photographs are taken with their edges parallel to the direction of flight (i.e. without *crabbing*) and with the fore and aft overlaps along each strip maintained at 60 per cent and the lateral overlaps between adjacent strips maintained at about 20 per cent.

We shall consider first the geometry of the single photograph which is assumed to be vertical. The *principal point* (PP) is the point where the lens axis meets the format at right angles; it is usually in the centre of the photograph and is defined by *collimation marks* which are either in the corners or at the midpoints of the sides of the photograph (see Fig. 7.1(c)). Sometimes the PP itself is marked by a small cross, but of course this is only possible when a glass register plate is fitted to the camera through which the film has been exposed. With suction back cameras the PP can only be marked on the photograph by joining up the opposite collimation marks. With a truly vertical photograph the PP will coincide with the *plumb point*, i.e. the image of the point on the ground vertically beneath the camera, but when the photograph is not vertical these points will not coincide and the distortions will then be radial from the plumb point. When they are close together, as in nominally vertical photographs, we assume that the distortions are radial from the principal point because this is easily defined while we do not know where the plumb point is.

Figures 7.1 and 7.2 should make three points clear. First, a tall object like a high-rise block of flats near the edge of the photograph is seen partly sideways and will obscure some of the ground on the side away from the PP. Secondly, the scale of any part of the photograph varies inversely as the depth below the camera and hence is affected by the height of the ground. Thirdly, as we have stated, the apparent displacement or *distortion* of the top of a tall object in the photograph away from its bottom (which is where it is shown on a map) is directly away from, or *radial* from, the PP of the photograph. This means that, even with a truly vertical air photograph, we cannot assume that the scale is constant or that the shapes and relative positions of features appearing on it will be the same as their true shapes and positions on the ground unless the area is flat. For example,

AIR PHOTOGRAPHS FOR PLANIMETRIC MAPPING

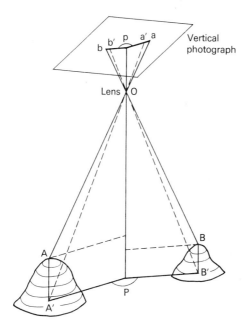

Fig. 7.2 (a) Radial distortion on vertical photographs. Note that $OPA'A$ and $Opa'a$ are in the same plane and $pa'a$ is parallel to PA', $OPB'B$ and $Opb'b$ are in the same plane and $pb'b$ is parallel to PB', and hence angle $A'PB'$ = angle $b'pa'$ = angle bpa.

it is clear that a straight road or hedge passing over a hill will only appear straight on the photograph if it lies under or points in the direction of the PP. If it runs at right angles to this direction and is distant from where the line through the PP meets the ground, then it will be curved away from the PP; its curve over the hill will be seen to some extent from the side and this will appear on the photograph.

This special property of vertical air photography, that the distortions are radial from the PP, holds well enough for all graphical work even in mountainous country and even if the photograph is not truly vertical but tilted by as much as 3°, which is about the limit to be expected for vertical photographs which have been taken properly. It leads to the fundamental property of such photographs that angles are true at the principal point but nowhere else (unless the photograph is truly vertical and the ground is flat). When tilts exceed about 3° (1/120) the positions of the plumb point (the image of the point vertically beneath the aircraft) and of the PP of the photograph obviously begin to separate appreciably (in this case by 6/20 in or about 7.5 mm) and the relationship becomes much more complicated. To draw a map from badly tilted or oblique air photographs requires specialized equipment and knowledge. However, it is worth mentioning that on a high oblique photograph of a flat area it is relatively easy to draw a perspective grid on the photograph, related to the horizon on it and corresponding to a rectangular grid on the ground. This method has been used in Canada with great success for small-scale mapping of large areas of tundra and lakes in the North West Territories.

The effect of this radial property of near verticals is the same as if in a field survey one was able to occupy successively the points on the ground vertically beneath the aircraft (the ground positions of the PPs) at the same time as each photograph was exposed. If one was equipped with a low-order theodolite, or an artillery director, or even a plane table, all of which can be used to measure horizontal angles with an accuracy of a few minutes of arc, one would then be able to measure angles from each of these points, but not of course from any others. The differences from a ground survey are first that rays can only be observed in this way to points contained within the same photograph and secondly that the points do not have to be visible on the ground from the PP. The surveyor turned cartographer can therefore measure not only the angles at each PP between the rays to the preceding and following PPs but also to suitable features in the corners of each photograph and at its sides (see Fig. 7.3(b)). For linking the photographs together these features, which are of no importance in themselves (e.g. they may be small bushes, termite mounds, or even cow pats (but not cows because they move) or anything that is reasonably small at the photographic scale, is not moving, and shows clearly on the photographs) must appear on all three consecutive photographs, i.e. they must lie on the narrow central strip (the *supra-lap*) onto which both the preceding and the succeeding photographs overlap. The surveyor is then running a double chain of triangulation in which he occupies successive stations on the centre line of the chain connected by forward and backward sights and intersects every point at both sides of it from three successive stations on the centre line but does not measure any distances (see Fig. 7.3(b)). This procedure establishes the *shape* of the chain completely and quite strongly, but does not of course fix its *size* or its position or orientation. These are done

AIR PHOTOGRAPHS FOR PLANIMETRIC MAPPING 173

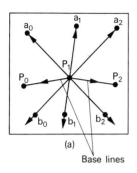
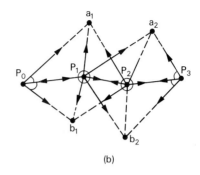

Fig. 7.3 The radial line principle: (a) radial observations from a single photograph; (b) the form of the aerotriangulation (measured angles are shown at P_0, P_1, P_2, P_3, the PPs of the photographs); P_0, P_2, images of preceding and succeeding PPs; a_1a_2, b_1b_2, pass or minor control points.

by including in the chain at least two ground survey points whose positions are already known and between which the *aero-triangulation* chain or strip has to be fitted. The other points of the chain which are called *pass* or *minor control points* are then fixed, and it is clear that they give a very dense net or framework of control on the plot being made from a strip of photographs and that detailed features can be interpolated between them in the way that all surveys should be made (Fig. 7.4).

When one strip has been scaled, oriented, and located at an appropriate scale (usually a rounded-off value of the mean scale of the photographs) in this way on a gridded plot of the area to be mapped, adjacent strips can be scaled and fixed in the same way by including in their aerotriangulation sufficient points of the aero-triangulation of the first strip lying on the common lateral overlap (Fig. 7.5). Ideally these should appear on the supralaps of both strips, but this is only possible if the photographs of the two strips are in phase, and normally they are not so they must be fixed by rays from the PPs of only two photographs of the second strip. However, if enough of these are used the connection should be strong, and therefore the scaling of the second strip will be nearly as good as that of the first. Successive strips can be scaled in the same way, and where they include new ground control points they are of course tied into these, thus correcting

174 AIR PHOTOGRAPHS FOR PLANIMETRIC MAPPING

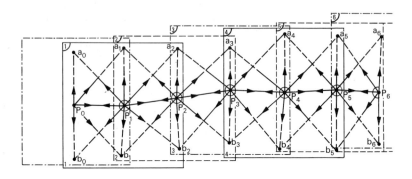

Fig. 7.4 Aerotriangulation of a single strip. The angles are only measured at P_1, P_2, etc. and only on the photograph of which each is the PP. Each PP, and each pass or minor control point, ($a_1 a_2$, $b_1 b_2$, etc), is transferred to the photographs immediately preceding and succeeding the one to which it belongs.

the small scaling and location errors which will have accumulated during the extension from the original strip. At the end of this process the whole area of the proposed map will be covered by rows of points of which alternate rows are the PPs of the photographs, and the rows between these are of the minor control points in the common lateral overlaps of the strips. At the PPs, but *nowhere else*, the angles are true and thus rays can be transferred from these to the plot in order to fix any point of detail appearing on the photographs which is to be mapped. Details of this process are given in the next section.

The minor control plot

For some unknown reason surveyors and cartographers use the terms *plot* and *plotting* for the work of drawing an original map from observations, as opposed to the fair drawing of the result which is printed; despite their sinister connotations we shall use these terms here for want of better. Compiling a minor control plot from air photographs comprises the following stages:

(i) laying out the right photographs on a gridded base;

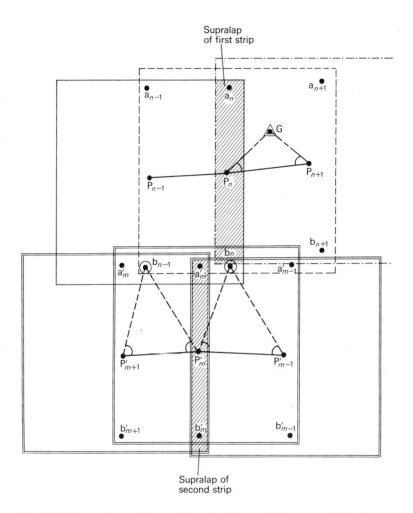

Fig. 7.5 Connection of two adjacent strips, which are not in phase, to each other and of one strip to ground control: P_n etc., PPs of first strip; P'_m etc., PPs of second strip; a_n, b_n, etc., minor control points of the first strip; a'_m, b'_m, etc., minor control points of the second strip; $G_\triangle$, ground control point on first strip.

(ii) selecting, transferring, and marking the ground and minor control points;
(iii) making and assembling transparent templates to represent the photographs;
(iv) drawing extra rays to fix detail points.

(i) *Laying out the right photographs on a gridded base*

As in planning a ground survey the first step is to decide on the limits of the area to be mapped, and then to find a flat surface large enough to accommodate this at the scale of the proposed plot. Normally, if the area is at all extensive, this will have to be the floor of a room, since a series of tables fixed together is likely to be unsatisfactory and not strong enough to take the weight of people crawling about on it as they will have to do to reach photographs in the middle. The scale should be close to that of the photographs. Alternatively the area can be divided up into smaller blocks provided that the framework of ground control is dense enough to fix the photographs in each single block satisfactorily. A single table may suffice if only a few photographs are being used for a relatively small area. The floor or table is covered with a white surface (paint or paper) and is then covered with a series of large sheets of transparent plastic taped together along their edges. A grid is drawn on this by what are in effect simple chain survey methods using large 3, 4, 5 triangles to obtain right angles. If done carefully with a steel or plastic tape this can be plotted to an accuracy of a millimetre. All the points of the ground control framework which have been identified on the photographs are plotted on this grid, or alternatively in a well-mapped area identifiable points of mapped detail such as house corners or path intersections can be plotted, or the existing maps can be enlarged up to the scale of the plot.

The photographs are then laid down on the plot in their approximately correct positions using the techniques described earlier and any redundant ones are eliminated, leaving a situation in which every part of the whole area is covered by at least two overlapping photographs.

If the photographs have been taken correctly each one will have a narrow area (the *supralap*) on it at right angles to the flight line in which three successive photographs overlap, and along their sides parallel to the flight lines these will also overlap laterally onto the photographs of adjoining strips. It is these areas of forward and lateral overlap which are the key to connecting the photographs together, and it is in these therefore that the minor control points must be located; time will be saved later if areas of lateral overlap are outlined in each photograph. All ground control points, whether surveyed and identified specially or selected from already mapped detail, are marked on all photographs on which they occur using a fine needle prick for the point itself and coloured pencils or ink for a triangle and a name or number.

(ii) *Selecting, transferring, and marking the minor control points*

A strip or run of photographs following one flight line is then chosen which has at least two ground control points on it, and if possible one where two of these are close enough together to appear on one overlap or at least on two successive ones. Each photograph of this strip is then laid out on a table with a good light facing the surveyor; and its PP is marked on it by joining up the collimation marks described earlier either by diagonals if they are in the corners or by joining up the marks at the midpoints of the sides. Only the intersecting parts of these lines should be drawn, using a hard pencil, and the resulting cross should then be pricked with a fine needle and ringed. The images of these PPs are then transferred to the preceding and succeeding photographs by putting each successive pair of photographs under a stereoscope, and they are also ringed and numbered. They should then be joined up on each photograph by lines radiating backwards and forwards from the photograph's own PP; these are known as *base lines*.

The middle parts of the lateral sides of each photograph are then inspected for small points of detail which also appear in the corners of the preceding and succeeding photographs. These can be any feature less than 1 mm in diameter which is recognizable with certainty on all three photographs, e.g. a small bush, a path intersection, or a manmade feature such as a small building or manhole cover. The ground corner of a large building can be used provided that it is not obscured by 'lean back' in one of the photographs which will often happen because it is the corners where the view is most oblique that are being used. These points are then pricked, ringed, and numbered, preferably by some system which relates them to the photographs where they are opposite the PP and not in the corners. The other strips can be treated in the same way until on every photograph there are six minor control points (and three PPs) which all also appear on two others. The minor control points can also be found on some photographs of the adjacent strips, but we shall leave this complication for the moment and concentrate on the treatment of one strip at a time, starting with the one that has the most ground control on it. To recapitulate, the nine points on each photograph will be its own PP, the transferred images of the preceding and following PPs, its own two minor control points opposite its PP, and the transferred images of the minor control points of the preceding and succeeding photographs in the four corners (see Figs. 7.3(a) and 7.3(b)).

(iii) *Making and assembling transparent templates*

We are now in a position to make a geometrical model or analogue of the photographs which uses the property that the angles at their centres are true. We choose the three or four successive photographs on which two ground control points appear and we make a transparent template of each by placing over it a stiff sheet of transparent plastic film the same size as the photograph and transferring to this the nine points (and any ground control points), and drawing rays from the centre PP to each of the other points with a fine hard pencil to create the diagram shown in Fig. 7.4. The points must be numbered as on the photographs. The templates are then assembled in order just as the photographs were, overlapping in their correct sequence, and the following two conditions must be observed;

(i) successive PPs must lie on the rays to them drawn on the preceding and succeeding templates;
(ii) the rays to each of the lateral minor control points from three consecutive templates must meet in a point, which is achieved by sliding the templates along the base lines joining their PPs until they do (Fig. 7.5).

If there are two ground control points in one or two consecutive overlaps it is not too difficult to preserve these two conditions while laying the assembly of three or four templates on the gridded plot and adjusting its position, orientation, and length so that all rays to the two ground control points pass over their positions on the plot. Once this is done the templates are correctly positioned, scaled, and oriented, and the vertical pins can be driven through the PPs and all the minor control points, both to hold these templates in position and to mark the correct positions of these minor control points on the plot. We are now in the same position as the plane tabler who, starting from two previously fixed main framework points, has begun to establish his own framework of minor control points from which to fix detail points in the features to be mapped. He can also extend its framework as a graphical triangulation until he can check its increasing inaccuracy by tying it in to another more accurate control point of the main framework.

Successive templates along the same strip can now be added to this small initial assembly following the two rules given in the preceding paragraph, but the new templates should not be pinned down until their positions have been checked, and probably slightly corrected, by their intersection with another ground control point, in the same way that any form of traverse is provisional until it closes on a fixed point of superior accuracy. When this strip of photographs and templates has been completed to cover all the ground control along it, adjacent strips can be connected to it using the now fixed minor control points along its lateral overlaps, just as if they were ground control points (Fig. 7.5). The images of these must of course be transferred to the photographs of the adjacent strips and thence to the templates representing these, and in fact this would normally be done when the minor control points for each strip are being chosen and transferred to all other photographs on which they appear. A complication occurs in that it is usually impossible for the aerial photographer to ensure that the overlaps of adjacent strips are in phase and opposite to each other (though this is sometimes achieved when good maps of the area are available), and so it is only occasionally that the supralaps of adjacent strips coincide so that the minor control point of one strip can be the same as that of the adjacent one. What normally happens is that the minor control point of one strip appears on only two photographs of the adjacent one and vice versa, but if enough of these are chosen on both strips the connection is very strong. However, as in the extension of the first strip from the initial assembly of three or four templates, these lateral extensions to adjacent strips must be regarded as provisional until they close on a further ground control point. Once the whole area has been covered and all ground control points have been used and have fitted correctly (see below), the assembly can be taken up and the position of each minor control point can be ringed and numbered on the plot.

(iv) *Drawing extra rays to fix detail points*

The plot is now covered with a series of rows of minor control points which appeared on the photographs and have been correctly fixed in position on the plot. These rows are composed alternately of the PPs of the photographs at which the angles are true, and the lateral points at which they are not, unless the ground is very flat and the photography has been very well taken. Unlike a framework established in the field the control is very regular since it is independent of the topography, but it is also rather inflexible since the positions of the points are very closely defined by the supralaps of the photography. Thus extra points will probably be required actually on or near the detail to be mapped in exactly the same way as the plane tabler, having established his own framework in places where it was easiest to do so, had to fix detail points from this on or near the features to be mapped. In the minor control plot this is done by locating these points on two or more photographs, transferring them to the templates, drawing rays to them on the templates, putting the templates down over the plot positions of their PPs, orienting them correctly on the existing minor control, and then pricking through the intersections of the rays to the new points required. If the ground is flat a number of such points can be established on each photograph and the detail can then be traced directly off the photograph between them or reduced using some simple optical device so as to fit between these points. If the area to be mapped is large and involves several photographs, it pays if possible to use two sets of prints, one for the minor control as described and the other for detail on which all the features to be mapped are interpreted and drawn in colour for transferring to the plot once the minor control has been established.

It will be clear that the procedure involves numbers of transparent templates and can be a complex and difficult operation. It is only practicable if there is a reasonable density of ground control of the order of one point to every two or three photographs; if the density of ground control is less than this then it is impossible to assemble the transparent templates and they must be replaced by *slotted templates* in which the rays are represented by accurately cut slots radiating from a small hole which represents the PP of each photograph. Studs with centre holes fit into this hole and into the slots, and a complete mechanical model is then made of the whole assembly in which the angles at the centre of each template are preserved unchanged and a long strip can be compressed or extended mechanically like a concertina (but still preserving its correct shape) to fit to control points which may be ten or more overlaps apart. Adjacent strips can also be adjusted laterally to fit other ground control points lying under different flight

AIR PHOTOGRAPHS FOR PLANIMETRIC MAPPING   181

lines. The use of this special equipment, the ancillary devices for plotting from a pair of photographs by means of the radial line principle, and the requirements for ground control, are described in the next section.

Radial line equipment and ground control

In the previous two sections the theory and practice (using only ordinary equipment) of establishing and using minor control points obtained by radial line methods from air photographs have been described. Once this has been done the transfer of detail from the photographs to the plot is straighforward since it has only to be interpolated within a close network of control points to which others actually on a feature can be added if desired. Obviously, since the air photographs give a map-like view and viewing a pair stereoscopically enables one to see a three-dimensional scale model of the ground and its features, this is a good deal easier than interpolating detail on the ground itself and can be done over longer distances. In many cases, with minor control points only a few inches apart, interpolation can be done directly by eye or by tracing off the photographs or, if a change of scale is involved, by projecting an image of the photograph at the right scale onto the plot and fitting it to the minor control points. This is particularly true when the ground is flat since the scale is likely to be reasonably constant over considerable portions of a photograph. However, it is clear that the procedures for establishing the minor control itself which use only transparent templates are really only suitable for quite small areas or, what comes to the same thing, areas with so much ground control that they can be divided up into small areas well controlled by a number of ground control points. However, two factors may make it advisable for special equipment to be used either for establishing the minor control plot or for interpolating the detail once this has been done. The first is if the ground control is sparse and the area large so that several photographs have to be connected together between adjacent control points; it then becomes impracticable to make them all follow the two rules given on p. 178 and also to fit over the control points. The solution, which incidentally was conceived by the nontechnical executive of an American air survey company watching his employees wrestling with several transparent templates, is to use slotted templates.

Slotted templates can only be made effectively by using a special cutter and punch which are shown in Figs. 7.6 (a) and 7.6 (c). The principle and operation are very simple since the basic idea is to replace the difficult graphical solution by a mechanical one in which the shape of the assembly is automatically preserved while its scale can be adjusted by sliding the templates towards or away from each other along the slots through their principal points, whereupon the studs representing the other points at the sides will then be forced towards or away from these in exactly the same proportions. Each template is prepared exactly as described on p. 178, and a centre hole is punched over its PP using a special punch with a spring-loaded central needle which locates the punch centrally over the pin prick representing the PP. The template is then placed in the slot-cutting machine with the PP hole over the stud provided. This slides in a groove which is a prolongation of the axis of the actual slot cutter which is aligned directly above it and parallel to it. The template is turned about it until one of the rays lies directly along the groove, and the stud is pushed along the groove until the point on this ray is under the slot cutter which is then brought down on to it. A final check is made by engaging a retractable needle in the cutter in the pin prick representing the minor control point through which the ray was drawn. The result is a slot radial from the PP following exactly the line of the ray. Similar slots are cut to replace the other rays and the template is then complete as shown in Fig. 7.6(b). Studs which exactly fit the centre hole and slide in the slots are fitted into them, and successive templates are assembled, as with the transparent ones, with one stud representing each of the three middle points on each template, i.e. its PP and the minor control points on each side of this. The slots in the preceding and following templates fit over these so that each PP fits into two slots and each minor control point into three slots (Fig. 7.6(d)). As the templates are slid towards or away from each other the whole assembly changes size but not shape, and when its scale is correct it can be fitted over the control points with only minor local adjustments which are evenly distributed throughout the whole. The studs have fine central holes (Fig. 7.6(e)) through which a needle can be pressed vertically to mark their final positions on the plot beneath once the whole assembly is complete.

The radial line plotter

The second problem mentioned on p. 181 arises when detail is to be plotted in an area so rugged that the features resemble their real plan

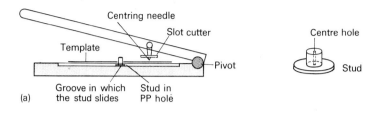

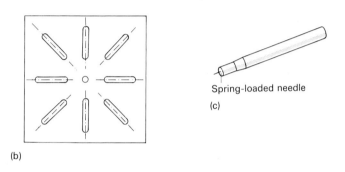

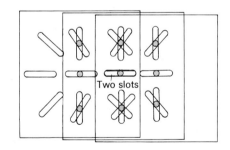

Fig. 7.6 Slotted templates: (a) slotted template cutter and stud; (b) single slotted template; (c) punch; (d) assembly of three templates.

shape on the ground only in a small central part of each photograph because of the height distortion away from the PP. Of course a large number of rays can be drawn to significant points on this detail on transparent slotted templates and then pricked through onto the plot, but this becomes tedious if the area to be mapped is at all large and it is better to use a mechanical device called a *radial line plotter*. A pair of overlapping photographs is placed on the tables of this device and adjusted so that the PPs and their images lie on the common base line but are separated so that they can be viewed in a stereoscope attached to the instrument. Transparent plastic arms pivot about the positions of the two PPs, and each has a fine line drawn on it which passes over the position of its PP. Although they do not in fact intersect, they appear to do so when each passes over a ground point which is viewed on both photographs, and a pencil linked to them by a parallel mechanism will then mark the position of the point on the map. The instrument includes an adjustable pantograph so that the scale of the plot can be different from that of the photographs, and so any feature appearing in both of them can be traced off by following it with the apparent intersection of the two radial arms. Anyone who has followed the principles of radial line plotting described above would be able to use the instrument after practising for a few minutes, and it is particularly useful for the kind of mapping required in forestry work in mountainous areas. Not only can the plan positions and shapes of streams and forest, soil, or land use boundaries, roads and tracks be traced off correctly despite their considerable changes in height and therefore in the local photographic scale, but the ridge lines can also be plotted, and when these and the stream lines are plotted on the map *form lines* can be sketched in to indicate the general shape of the topography, although of course without putting in any heights (see also Chapter 11).

Ground control for radial line plots

As already described, use of the radial line principle with either transparent or slotted templates ensures that the photographs and the minor control points on them — and any extra detail points fixed radially — are in their correct relative positions and that the framework thus created is the correct *shape*; what it cannot do alone is to give the correct *size* or scale, or orient and locate it on the earth's surface. The only way this can be done is to identify on the photographs, and draw rays or cut slots towards them on the templates, a number of points whose positions are fixed on the ground by field survey – the *ground control*. The correct and accurate identification of these points on the photographs is considerably more difficult and liable to error than most surveyors and other users of air photographs tend to realize, and a separate section is devoted to this problem in Chapter 9. In this section we are concerned only with the choice, number, positioning, and use of such points for the planimetric control of a radial line type of plot, i.e. one in which no heights or contours are shown.

The minimum number of ground points required to control a slotted template assembly is clearly two, and if they are reasonably far apart they will scale, orient, and locate the whole assembly, although outside the line joining them and in its outer margins the assembly will be relatively loose and will therefore accumulate errors of scale and position. However, it is a major advantage of this technique that a surprisingly accurate uncontoured map can be made from a large number of air photographs of a large area using a minimum of ground control; in fact millions of square miles of relatively uninhabited areas of Africa have been mapped by agencies like the French Institut Geographique Nationale, the Directorate of Overseas Surveys, and the United States Geological Survey using this technique and either vertical or oblique (on which the technique is more complicated) air photographs. Clearly more ground control points are required for more accurate work, and the sort of density often quoted is a minimum of one control point to every seven photographs along a strip or transversely one point in every third or fourth strip.

The most effective location for the points, if this kind of density cannot be achieved, is round the perimeter of the assembly since this is where the templates are held only at one side and are therefore most liable to movement and distortion. In the middle they are so closely and tightly interlocked that they have very little freedom to take up any except the correct positions. In relation to the photographs themselves the one location to be avoided is clearly on or near the base line joining two successive PPs since the rays or slots will then be in line and there is no proper intersection of the point or stud. The best location is out to one side and in the supralap of three successive photographs; if the point also lies on two (or even better three) photographs of the adjoining strip it will clearly give a very tight control over them. It is clear from this that the actual points used to control the photographs are best chosen *after* the photography has been taken rather than before, even though the actual field survey framework on which their positions are based may have been established before photography. One further point about location is worth remembering, and this is that it is a great convenience to the person assembling the templates, as mentioned on p. 178, if he can find two close ground control points either on one overlap or at least on two successive ones so that before assembling a larger number of templates at possibly the wrong scale he can start with a small initial assembly of templates which are correctly scaled, to which others can then be added. It is of course important that both of these initial points should be completely reliable both in their co-ordinates and in their identifications.

The methods used for fixing the ground positions of these points are exactly the same as those used in fixing framework or detail points in field surveying, i.e. chain survey, triangulation, traverse, bearing and distance, or by plane table; but their locations are dictated by the requirements of photography rather than of convenience in field survey. They must also be on easily and reliably identified sharp objects rather than on the tops of hills with a good view, in addition to complying with the rules about their positions on the photographs given above. Very often this means that they are, so to speak, appendages of the framework rather than actual points in it, and they are often fixed from framework points by such methods as bearing and distance in which gross errors can occur. It is therefore particularly important to include and record full checks on such fixes, for example by measuring all distances in both feet and metres and all angles and bearings with two completely different settings of the theodolite. In the sort of rugged forested areas found in some tropical countries the only points which can be identified on the air photographs will be in clearings along roads or rivers, and these may often be some distance from the nearest trigonometric point on a hill top, so particular care is then required to avoid errors, preferably by running closed traverses through them.

Where good existing maps are available control can be taken from these following the same rules for identification and location in relation to the perimeter of the area and to the photographs. The most likely points are man-made features or hard detail such as small buildings, bridges, road or track junctions, pylons, and so on; where the map and photographic scales are large, particular features on these such as house corners, the ends of parapets, corners of paved surfaces, and so on will have to be chosen; but in all cases care must be taken that there is no gross misidentification (e.g. the wrong house or road junction) and that no change has occurred between the date of the map and that of the photographs. The DOS 1/50 000 and 1/100 000 scale maps usually include the plotted positions of the PPs of the photographs from which they were compiled. If these photographs are being used by the scientist, then no further assembly is required; he can plot the detail he requires directly from the PPs as described earlier (p. 180). However, if he is using more recent photographs he will have to select points of detail appearing on both sets, make a radial line plot from the *original* map photography, and then use the points of this as if they were ground control from which he can fix a radial line plot from the *new* photography. This of course means obtaining prints of the original map photography.

8. USING STEREOSCOPIC PAIRS OF PHOTOGRAPHS

Introduction

In the previous chapter we described the use of air photographs for planimetric mapping only (i.e. without contours) by techniques which assumed that all the photographs were truly vertical ones and by handling them one at a time and not in pairs. Stereoscopic pairs were only used for transferring the PPs and points of detail from one photograph to the adjoining one and for sketching in features between them. These techniques are approximate, they do not measure heights or slopes, and they can only be used accurately where the scale of the map is near to or less than that of the photography. As already mentioned in Chapter 7, one of the two reasons why aerial survey photographs are taken with successive overlaps of at least 60 per cent and with a good overlap between adjacent strips is to ensure that every part of the area is covered by at least two photographs so that any part of it can be viewed stereoscopically. In this chapter we first describe how the stereoscopic effect is produced and then consider the simplest uses of stereoscopic pairs of air photographs which are (i) the interpretation of ground features and (ii) the measurement of relative heights (e.g. of trees). The more complicated uses of stereo photographs, i.e. (iii) accurate three-dimensional mapping and (iv) the production and use of *ortho-photographs*, are discussed in Chapter 9. Ground-control requirements for accurate three-dimensional mapping and the reverse operation of using this to provide a framework for the scientists carrying out detailed mapping on the ground of features not visible in the photographs are also described.

Clearly, (i) and (ii) are directly within the capability of an agricultural or other field scientist, while (iii) and (iv) are not, mainly because they require elaborate and expensive instruments and special skills. Nevertheless, the scientist may require to use or commission surveys requiring techniques beyond his own capability, and so he must have some understanding of these more advanced techniques so that he can appreciate their weaknesses and their strengths, realize what is possible and also what is not, and have some idea of whether a contractor is overstating the accuracy of his proposed or completed work. He may also, by close co-operation with photogrammetrists, be able to save himself a good deal of unnecessary field work.

We have seen in Chapter 3 that of the three dimensions length, breadth, and height, the last is by far the most sensitive; and on the ground it is often measured to a much higher degree of accuracy and far more easily (although tediously) than the other two. Even in the earliest civilizations canals were being dug over long distances with extreme precision in their levels, although their positions on the ground were recorded, if at all, only on very rough maps. Field survey techniques have carried on this tendency, and often determine position and height as two separate operations and with the latter measured to a much higher degree of accuracy than the former. Any competent field scientist equipped with the right instrument can level to an accuracy of 10 mm per kilometre but few could achieve this accuracy of position. It is important to realize that in the case of air photographs and of maps made from them the reverse tends to be true, and that there are two fundamental differences between maps made from ground surveys and those made from air photographs using three-dimensional plotting machines. First of all, the ground surveyor never actually traces the shape of a feature; he establishes a series of detail points and interpolates its shape between them. Secondly he either calculates heights from small vertical angles and measured distances or he measures differences in height directly using a spirit level or an aneroid barometer. In all cases except the last the result is considerably more accurate than the horizontal distances or positions.

The photogrammetrist plotting a line map from stereoscopic pairs of photographs works quite differently. First, he does not have to fix a series of points on each feature and then sketch between them; he traces the feature directly. Secondly, since he is working with a three-dimensional model, his height measurements are no more accurate than his horizontal ones; in fact there are technical reasons why they may actually be less accurate. His immediate area of interest is not a field sheet or a single photograph; it is (as described on p. 162) the overlap between two successive photographs in the same strip. This overlap is a near rectangle defined by *neat lines* which is approximately 40 per cent of a single photograph along the flight line and 75 per cent at right angles to this. Its corners are formed by the pass or minor control points already described in Chapter 7, and these may have been fixed directly by ground survey, by slotted template, or by the more elaborate, accurate, and three-dimensional techniques of aero-triangulation, based on more widely spaced ground controls.

Having established this network of minor control and divided the area into separate overlaps which meet along their neat lines, the photogrammetrist then sets up the stereoscopic pair forming each overlap in a three-dimensional plotting instrument which exactly reproduces the conditions under which the two photographs were taken, except that it scales down the distance between the two points at which they were taken from its true value in the air to what it would be on the map which he is making.† In effect, like the plane tabler, he resects the positions of the two camera stations in the air from the minor control points as if they are a pair of towers from which he is viewing the ground beneath. If he were a ground surveyor he would visit each air station in turn and draw intersecting rays to individual points in the overlap and interpolate between them. Now he views the overlap stereoscopically and sees a three-dimensional *model* of the ground, which is positioned, scaled, and levelled by bringing it down onto the four minor control points at the corners. Three would of course be enough, but like all good surveyors he wants a check, and this is provided by the fourth; in any case it is obviously easier and more reliable to work from a complete series of corner points than from only three of them. On the scale of the photograph the overlap will be about 7 in × 3.5 in (170 cm × 90 cm), and within this area his plotting will be everywhere of equal accuracy provided of course that he can see the features and that they are not obscured by trees, masked by buildings or overhangs, or covered by vegetation. As we have seen on p. 169 the accuracy and resolution of good air photographs approaches 5–10 μm on the scale of the photograph, and so in theory they can be enlarged 50 times before errors exceeding 0.5 mm will appear at plotting scale. In fact enlargement by a factor of 10 is the absolute practical limit and it is more common to map at not more than six or eight times the scale of the photograph. Even for this degree of enlargement elaborate and expensive instruments are required, but the advantages are considerable. At this scale of enlargement each overlap will be about the size of a plane table sheet with a control point at each corner, and over this area the photogrammetrist can trace every detail he can see with plottable accuracy and without having to move from his comfortable office or having to take any more instrumental measurements. Comparing this with the

† The suggestion in White (1977), p. 23, paragraph 4, that the photographs have the effects of tilts removed before being placed in the plotting machine is incorrect.

problems facing a field surveyor working in the same area, we can see at once why aerial survey has taken over even at the very large scales used in towns and in some engineering surveys. We can also see why, for topographic mapping, photogrammetrists like to use photography on as small a scale as possible without it being so small that man-made features cannot be interpreted correctly.

Because the model is three dimensional, the tracing device must also be movable in three dimensions; it consists of a small *floating spot* which can be moved not only along and across the overlap but also in and out of the model, and, like the two photographs, it is also being viewed stereoscopically. When tracing a feature the spot must always be made to touch the ground, or with buildings a feature like a roof edge which is known to be vertically above the ground plan of the building. Provided that this is done, the pencil of the instrument which is related to the spot will draw the correct ground plan of the feature on the plot. If ground heights are required the spot can be stopped and brought down carefully until it is exactly onto the ground; the height is then read off on an altitude scale. To draw contours the height scale is locked at the contour value; by horizontal movement only the spot is then made to touch the model everywhere at this height and the plotting pencil traces out the contour. We can see from this how different the process is from that of ground surveying. First, detail mapping is a continuous process in three dimensional photogrammetry; features and contours are traced directly and continuously by methods which are almost as accurate (except that the spot is moving) as the fixing and heighting of individual points. This means that the photogrammetrist can record irregular natural features such as stream beds including every minor curve and change of direction, whereas the ground surveyor inevitably has to smooth and round off the feature between his detail points, particularly of course if he is at the same level and unable to view it from above. Maps of natural features made from air photographs always look much more detailed than those made on the ground, and the same applies to the contours.

However, as we have seen, in three-dimensional photogrammetry height measurement is no more accurate than horizontal measuremnt, and may be less so. As a general working rule 95 per cent of spot heights can be established to an accuracy of something like 1 part in 2000 of the flying height above the ground, and 95 per cent of contours (with a planimetric accuracy of half the contour interval) to

1 part in 1000 of the flying height. Therefore 1/10 000 scale photography taken from 5000 ft can be used to produce 5 ft contours. This assumes good ground visibility and good photography in which the tilts have been kept down to a few degrees, and that the photographs are free from distortion due to camera or film defects. These conditions are not always realized and it is then that the weakness of photogrammetric mapping appears: the contours may be the right shape and show every minor change of slope, but they may be displaced horizontally on the map to give a false value of the absolute heights in a particular area. A particular cause of this is when the film has not been truly flat during exposure. It is clear that photogrammetry is most useful in mapping rugged and inaccessible mountain slopes where exact heights are less important than a picture of their general shape and where ground survey would be very slow and difficult; and is least effective in mapping very flat areas where contour intervals of less than 1 m may be required so that absolute height accuracy is essential, and where minor irregularities like grass tussocks or termite hills are unimportant. When commissioning maps of areas being considered for irrigation the agriculturalist should be most cautious in accepting claims by the less scrupulous air survey operators, particularly if the photography has been or will be taken when the ground is covered by crops or other vegetation. In this case the right answer may well be a map of which the planimetry is drawn from relatively small scale aerial photographs but the contours are interpolated between a close grid of heights established by spirit levelling on the ground. Because this is tedious, slow, and expensive, especially overseas, there may be a temptation to accept a less accurate or coarser result produced by photogrammetry; but at present there is no real substitute for slogging it out on the ground.

All this explains the four uses of stereoscopic pairs listed on p. 188 and described in the following sections. In the first two we are only looking at the general shape of the ground over each overlap, or, if viewing individual features in more detail, we are only using the third dimension to see their shapes as a help to identifying them or measuring their heights relative to detail close by. No attempt is made to set up a complete and accurate model; we are still assuming that the photographs are true verticals. This is a very different matter from establishing absolute heights over the whole overlap or series of overlaps related to ground control. In the next chapter we describe how accurately positioned, scaled, and levelled models of the individual overlaps are created and then used for drawing uncontoured maps or plans at much larger scales, for drawing contoured maps at any scale, for establishing a framework of fixed points or features to help the field scientist, or for rectifying it or making an *orthophotograph*, both of which produce a copy of the photograph in which all the distortions and scale changes due to camera tilts or changes of height (whether of the camera or the ground) have been removed. This can only be done accurately be rectifying single photographs when the ground is flat, but in hilly areas it can be done by setting up accurate stereoscopic models exactly as for line mapping. The photography of each overlap can then be rectified and reproduced as a single accurately scaled map over the whole overlap area, provided that the slopes are not too steep. If contours are required they can be plotted in exactly the same way as on a line map or they can be interpolated between a series of profiles. However, before we describe these operations in more detail we have to discuss the principles of stereoscopic vision and how it operates, since we are now talking about a method of measuring distances which is quite different from anything we have described so far, involving the use of both eyes simultaneously and a mental process which is still not fully understood.

The stereoscopic principle

How in fact does an observer measure distances using only his eyes and brain and no instruments? The answer is that he uses three different techniques.

(i) He appreciates the amount of 'squint' or convergence needed to look exactly at the same object or point on it; this is the stereoscopic effect.
(ii) He appreciates the changes in the shape of the eye lens required to focus on the object. A recent television programme showed that chameleons use the focusing of only a single eye for measuring distance. If fitted with spectacles which alter this they fail to catch insects because they misjudge the distances.
(iii) He appreciates the angle subtended at the eyes by an object of known size by measuring the size of its image on the retina.

We are concerned mainly with the first of these but messages are sent to the brain by all three, and if they conflict this will cause problems which have to be overcome. The first two in particular are closely related from babyhood in the same way that a cyclist relates balance and steering; he has to break this association when mounted on a tricycle — and usually follows the road camber straight into the ditch at his first attempt because he does not! In learning to use a pair of pictures to achieve a three-dimensional model we have to break the stereoscopic focus association, and although various devices are available to help us (and some people can manage without), some people with apparently normal sight are unable to do so. The importance of the stereoscopic technique in precise and rapid measurement of distance and appreciation of shape is shown by its evolution in all the larger predators (including birds) and in the apes, monkeys, and man who required it for their more precise and detailed handling of food and tools. The fact that more primitive forms of animals (and one-eyed people) can live successful lives shows the importance of the other techniques as alternatives.

Although it depends on the measurement of a remote angle by intersection from a fixed base, stereoscopic measurement does not use the same functions of eye and brain as the range finder. The range finder receives two images of the distant object through two objectives a fixed distance apart, but these are then viewed by *one* eye and made to coincide by a mechanism which measures the angle subtended by this distance at the object and either records the distance on a scale viewed by the other eye or (in a camera) focuses the lens on the object being viewed. Only one eye is used. In the normal stereoscopic vision of real objects the brain measures distances by *fusing* the two slightly different pictures of the object received by the two eyes. The brain measures its *absolute* distance away and also appreciates its shape by looking at different parts of it to measure *relative* distances (Fig. 8.1). This fusing process consists of the rapid simultaneous scanning of a small area at a time by both eyes and the matching of the image patterns of colour and shade received rather than looking fixedly at a single point, and some automatic stereoscopic plotting instruments also use this technique. It will be clear from this that certain conditions must be fulfilled if successful and accurate stereoscopic measurement is to be achieved.

(i) The ratio of distance to viewing base must be neither too great nor too small.
(ii) Both eyes must be looking at the same object (or images of it), and this must either be small and definite enough or have

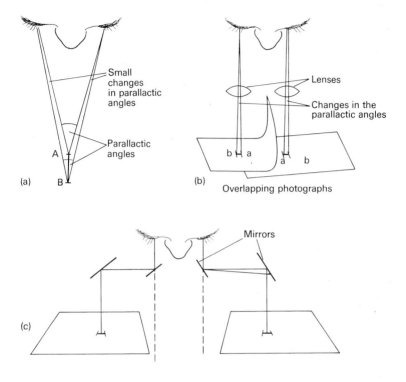

Fig. 8.1 The theory of stereoscopic vision: (a) Normal stereovision: eyes squinting and appreciating absolute and relative distances or heights of A and B. (b) Stereovision on aerial photographs: eyes parallel but seeing changes in the parallactic angles of the top and bottom of the tall object AB. Simple stereoscopic lenses increase the apparent focusing distance to near infinity. (c) The mirror stereoscope makes physical separation of the photographs possible and also increases the viewing and focusing distance to correspond more nearly to the apparent stereoscopic distance.

enough texture for both images to be matched in the fusing process.

(iii) Both eyes and the brain must be capable of producing and handling this matching process even if the apparent distance conflicts with other information being received.

(iv) Where the stereoscopic effect is produced by two photographs not taken simultaneously (and they never are in aerial photography), there must be no relative movement of the objects or of their images being viewed during the interval between the two expsoures.

The base to distance ratio

Beyond moderate distances the normal human eye base of about 2.5 in (60–65 mm) provides no stereoscopic appreciation of distance. For example, from a point at sea level it is impossible to tell which of two ships about the same distance out to sea is the further off until they pass each other. The same is true of aircraft when looking across an airfield, and this principle is used by formation flying teams to shock their spectators by flying two aircraft in opposite directions across the field at low level and apparently on a collision course. Looking down from a flying altitude of only a few thousand feet even quite hilly ground appears flat because the 'squint' or *parallactic angle* at the ground, and changes in it, are then very small. For example, at 5000 ft the base-to-altitude ratio is only 2.5 to 60 000 in or 1 part in 24 000 or about 8″ which is near the resolving power of the human eye. Small differences in ground height produce even smaller changes and unless other information — such as shadows from a low sun or knowledge of the shape and heights of trees or buildings — is available, no appreciation of height is possible. In stereoscopic photography this is overcome by separating the two camera stations and looking at each photograph with only one eye, by using a lens with as wide an angle as possible to produce the maximum effect. The observer is then turned into a giant with an eye base equal to that of the distance flown by the photographic aircraft between successive exposures — or alternatively the camera reduces the ground scale to such an extent that its depth below the observer is only about one and a half times that of his eye base when using a wide-angle lens. However, to view this model directly the giant would then have to squint horribly (and even more so with a superwide-angle lens where the depth to base ratio is unity). Try fusing an upright pencil as close as possible to the eyes and then measure the distance with a scale; it will not be much less than 3.5 in (90 cm). We show below how this problem is solved.

There is another limitation at the larger end of the base/distance scale since the differences between the two images of very steep slopes or tall buildings may be too great to be accommodated. Lenses of longer focus may have to be used for the photography of towns or very rugged country, or stereoscopic viewing may have to be restricted to the middle section of the photographs which then must be taken with a much larger overlap than the normal 60 per cent forward and 25 per cent laterally. Lenses of longer focus can be used to achieve a strong stereoscopic effect from a short base by the use of convergent photography, or in the laboratory by using a single camera but rotating the object between the two exposures.

Matching the images

Anyone who has picked gooseberries, which are usually similar in shape and size to each other, may have suffered from trying to pick a *pseudoscopic* one,† when the eyes have seen two different fruit through the interstices of the bush and the brain has fused them into one — of course at a fictitious distance and direction. These are clearly defined but misidentified objects. Lack of texture and definition with which to fuse the two images is the reason why it is quite difficult to fix a peg to the modern smooth plastic-covered clothes line. The solution is to put the head on one side and the eye base at right angles to the line — provided you do not mind what the neighbours think! For the same reason man-made features (provided that they are not too repetitive and the same ones are being viewed on each photograph) or trees and bushes are relatively easy to fuse on a pair of photographs, but featureless desert or snow fields will cause trouble unless either the sun is low enough to give shadows of water courses or wind-blown features or there are patterns caused by vegetation, or gravel and sand patches. Since many agricultural projects are planned in flat semi-desert areas, this is important because these may be not only where the most accurate and closest contouring is required but also where the photogrammetrist experiences the greatest difficulties in doing so. It will be clear that in a three-dimensional plotting instrument, while there is no difficulty in fusing the precisely defined images of the floating dot, unless the ground on which it has to be placed has a good texture this cannot be fused precisely and so the dot cannot be placed precisely in contact with it.

† Photogrammetrists use this term only for a false image produced by reversing the photographs so that heights appear as depths and vice versa, but we have used a wider meaning of the term here in order to introduce an attractive and, we hope, helpful concept.

Stereoscopic ability or acuity

The human eyes at rest tend to look into the middle distance; their axes are parallel and their lenses are focused to a distance of at least several feet. Close focusing and stereoscopic fusion require an effort and this is why two classic symptoms of drunkenness or extreme fatigue are blurred and double vision of any relatively near object because this effort cannot be made. Clearly no-one would try to work with air photographs when drunk, but fatigue may have the same effect, and clearly also, as we shall see below, for prolonged work the observer's eyes must be used in this most restful state and not at the large angle of squint required by direct viewing or at very close focus. This is achieved by interposing some combination of lenses and prisms or mirrors (the *stereoscope*) between the eyes and the photographs, but even then some people cannot achieve stereoscopic fusion. It seems that a surprisingly high proportion of people with otherwise adequate and well-balanced eyesight — about 1 person in 20 — have this disability. Only carefully designed tests can discover this weakness, and all organizations engaged in photogrammetric mapping apply these both to recruits for their photogrammetrists and to field surveyors. The latter will be establishing ground control, including the essential identification on the air photographs of the features fixed by ground survey, and good stereoscopic vision is necessary for this. Anyone planning to use aerial photographs would find it worth while to undergo such a test. The tests should involve the use of pairs of photographs with good texture but no marked shadows, and also of test cards in which pairs of dots or small circles are fused and where some are slightly displaced so that they should appear above and not on the surface of the card. The candidate has to identify these. However, those demonstrating the use of stereoscopy to distinguished visitors with no previous stereoscopic experience are advised to supply a pair of photographs of easily recognized buildings with well-marked shadows falling towards the observer who will then appear to see a solid model without necessarily fusing the two.

The reader may like to test himself by making his own card. First look in the mirror with a scale in front of the eyes and measure their distance apart. Normally this will be between 6 and 7 cm. Draw or type on a piece of card two identical dots or circles about 35 cm apart, and above them two others set inwards at about 2 mm each, as shown in Fig. 8.2. Then hold a piece of stiff card vertically between

Fig. 8.2. Test for stereoscopic vision.

them so that each dot can only be seen with one eye, and look down on the dots from a sufficient height so that they can be focused, using reading spectacles if necessary. If you are capable of stereoscopic vision it should then be possible to fuse both dots so that only one of each is seen, but the pair which are closer together will appear to be nearer to you than the other. The illusion can be helped by using circles and joining them by lines as shown so as to give a picture of a tall thin cylinder, which is how a factory chimney would appear in a stereoscopic pair of air photographs of it. If a hand stereoscope is available the dots or circles can be drawn further apart and at a distance more nearly equal to the eye base, since the focusing/stereoscopic conflict will then be removed by the lenses of the stereoscope. It is clear from this that if we can arrange for two dots to be viewed, one by each eye at the same time as we are viewing two corresponding areas of a stereoscopic pair of air photographs, and if we can change the distance between the two dots, then we can create a *floating spot* whose single fused image will appear to move vertically relative to the model which we see when looking down at the two photographs. If the distance between the two dots is exactly equal to the distance between two identical points of detail on the photographs the fused floating spot will then appear to be in contact with this detail; if the two dots are closer than this it will seem to be above the detail and if they are further apart it will appear to be below this. This is how a floating spot works, and we shall see later how it is used both in the parallax bar and in three-dimensional plotting machines.

Movement

In normal vision both eyes receive simultaneous signals and therefore no movement of the object can occur between their reception at each eye, but clearly with aerial photography in which there is a time interval between exposures this can very easily occur. People, animals, or motor cars which have moved small distances parallel to the flight line between successive exposures appear to be either flying or underground and are easily discounted; but a more serious (because less easily detected) effect is sometimes caused by the waves in crops or grass caused by a strong wind; low-level photography taken for close contouring will be completely useless in such conditions.† Movement of the camera during the exposure also causes inaccuracy; if a *focal plane shutter* with a travelling slit is used then the complete exposure may take a tenth of a second, during which the camera will, at a speed of 200 miles h^{-1}, have moved nearly 30 ft or 10 m. Reconnaissance aerial cameras such as the old RAF F24 used focal plane shutters, and photographs taken with such cameras can only be used for small-scale work without contours. Satellite pictures are also taken by a scanning mechanism which, at the speeds they are travelling, spreads the exposure over a considerable movement of the satellite, but because of their very small scale these pictures cannot in any case be used for contouring at contour intervals of less than about 1 km and for planimetric work the copies which are issued have usually been rectified from the accurately known orbit of the satellite.

The practical application of the stereoscopic principle to the measurement of heights in air photographs is not quite as simple as it appears in many textbooks. This is because there is an important difference between what happens in the normal stereoscopic vision of solid objects and what happens when viewing a stereoscopic pair of photographs. In the first the eyes are, so to speak, looking through the lenses of two cameras at the same object and recording the slightly different images on their retinas. In the second they are looking at the two images already recorded by the same camera at two different points in space, and the geometrical arrangements of the camera stations and of the eyes viewing these images are not identical. In real life both eyes looking at the same object use both the absolute and relative parallactic angles to measure both its distance away and its shape or its relationship to objects near it. However, as is shown in Fig. 8.1(b), in order to use the eyes in their most restful condition when viewing a pair of air photographs, these are viewed with the eyes looking almost parallel and thus they are recording an apparently large absolute distance. In contrast, with photographs taken with wide-angle lenses, the changes in the parallactic angle are very large

†Water surfaces cannot be fused because if calm there is nothing to fuse, whilst waves or floating objects must be assumed to be moving. Sometimes the water's edge is used as a reference level, but care is required because the apparent edge represents the limits of penetration by rays of light and this varies immensely with the lighting, the angle of incidence, the purity of water, and the type of film being used.

for the quite small relative differences in distance which reveal its shape as if the object were only a few inches away. Two normal relationships are thereby upset: the stereoscopic/focus relationship in the measurements of distance as already described and the normal relationship between the absolute and the relative measurements of distance assessed from the parallactic angle and from changes in it. The first is accommodated either by introducing lenses which appear to increase the focusing distance to something nearer the apparent stereoscopic distance (as well as magnifying the images), or by lengthening the light path between the eyes and the photographs by the use of mirrors and prisms (Fig. 8.1(c)). With experience it is even possible to achieve fusion by holding the photographs at arm's length, which has the same effect.

The accommodation of the second change of relationship, between the absolute and the relative stereoscopic distances, has to be achieved in the brain. This is confronted on the one hand by the more distant focusing achieved by the stereoscope and the near parallelism of the eye axes, which both indicate a distant object, and on the other by the very large changes in parallactic angle which occur as the eyes scan a relatively small part of the object and make it seem very close. It may be that it is the failure to accommodate these conflicting messages that leads to a lack of stereoscopic ability in people with otherwise good and well-matched eyes. The initial effect on the normal observer is to make objects in a stereoscopic model appear very much taller than they really are, but with experience this exaggeration is soon discounted and he uses his greater sensitivity to height differences to appreciate and measure their real shapes more accurately. The reasons for not viewing the photographs directly through apertures or lenses exactly replacing the positions of the camera at its two stations are not only the one we have seen, i.e. to let the eyes operate in a restful state, but also the physical difficulty of overlapping the two photographs and the need to separate them. What is clear from all this is that while the observer can have a very sensitive appreciation of *relative* height differences over a small part of the photographs, because of the large parallax changes he is quite unable with just a stereoscope to make any *absolute* measurements of height or depth over the whole overlap. The ordinary air photograph interpreter can estimate relative heights (and so recognize objects) quite accurately and can actually measure local height differences with a parallax bar costing a few pounds, but it requires an elaborate optical, optical-mechanical, or electronic instrument to reconstruct the geometry of the two air photographs in order to be able to measure absolute heights over a whole overlap or any large section of it. Here also the eyes appreciate small height differences very accurately; but it is the geometry of the plotting instrument and its reconstruction of the original geometry of the two camera stations which alone make absolute height measurement possible.

Interpretation

It is a basic axiom of photographic interpretation that there is no profession of photographic interpreter; anyone interpreting aerial photographs must be familiar in real life with the objects seen in the photographs whose images he is viewing and interpreting into real objects. This book is concerned mainly with measurement, and in these chapters with measurements derived from air photographs, so it will give no detailed guidance to agricultural or other scientists on how to interpret what they see, except for survey and measurement. However, a few simple points are worth mentioning. Clearly the principal difference in relating the use of air photographs to ground experience is in the point of view; with vertical air photographs one is looking down while on the ground one is usually looking horizontally, seldom downward, and never vertically. This is why looking at single vertical photographs is usually a profitless occupation; without the perception of height misleading impressions may be gained. An extreme example of this occurred to me when a cartographer being trained to use air photographs showed me a 'huge building' in one photograph of an irrigation scheme which was in fact one rectangular field flanked by two others shaped like parallelograms, the three together giving the impression of the top and two nearer sides of a very large block. A moment's look at the stereoscopic pair, instead of only one photograph, showed that the whole area was quite flat and that the 'building' was an illusion.

Suppose that you have been given a pair of photographs and asked to look at them stereoscopically and interpret the features on them. The first step is to set them out correclty. The PPs of the photographs should be transferred and the base lines drawn as described on p. 177, the photographs should then be set out with the two base lines parallel and on the same line, which should itself be parallel to the base of the stereoscope. The common overlap should be on the inner half of each photograph. If a mirror stereoscope is used the photographs should be set with their PPs under the centres of the two outer mirrors, or slightly closer together than this. A recognizable point of detail should be selected on each photograph and the two brought together along the base line — if necessary be placing a finger over each to locate them more easily — until fusion is achieved. The photographs should then be taped down lightly and the stereoscope moved so as to scan the overlap and check that fusion can be achieved everywhere, the photographs being adjusted by small amounts if necessary to obtain this. They should then be stuck down firmly with tape or pins.

So far we have described only the use of the paper prints normally available in the field and in the office. However, a considerable proportion of the detail available on a modern high-resolution aerial negative is lost on these. The surveyor/photogrammetrist only uses paper prints for relatively low-order work, for graphical techniques, or in some approximate plotting machines. Glass or film *diapositives* should be used to extract the full value of the photographs. These cost more, but this extra expense and the extra cost of the equipment required for handling them can often be far outweighed by the expense saved either in not taking photographs on an unnecessarily large scale or in not requiring special photography where existing cover will be satisfactory with the right viewing techniques. The greater detail available in glass or film *diapositives* (which can be placed between glass plates for viewing) is due to the finer grain of the emulsions used on film and glass compared with paper and also to the use of transmitted instead of reflected light. Stereoscopes designed to use diapositives and transmitted light are already available to photogrammetrists. Not only can the extra cost of this equipment and of the diapositives be covered by savings in the cost of taking and controlling photography if this is at a smaller scale, but it will be found, as the photogrammetrist has long known, that the handling of smaller numbers of photographs produces its own economies as well as increasing the accuracy with which they and the features on them can be positioned. In general soil surveyors and geologists do not require photography at much larger scales than that used by surveyors and cartographers, but foresters are very much inclined to ask for photography at a scale of about 1/25 000 when 1/40 000 or 1/50 000 will be adequate for all other users. This will either require special photography which is very expensive, or making other users handle photography which is at a larger and less economic scale than they themselves want. It is by no means clear whether foresters have examined the possibility of better viewing techniques at higher magnifications which would make it possible for them to obtain just as much information from photographs normally used by surveyors, photogrammetrists, and other earth scientists as they do at present from paper prints at larger scales.

Measuring the heights of objects

As explained above on p. 202 it is virtually impossible and certainly impracticable to measure even relative heights between widely spaced points in a whole overlap without the use of a properly designed three-dimensional plotting instrument which reproduces the geometry of the two cameras and their relative positions, heights, and attitudes, either physically in an analogue instrument or by elaborate computations in an analytical one. However, relative heights over small areas can be measured using a quite cheap and small instrument known as the *parallax bar* (Fig. 8.4). Without a three-dimensional instrument we cannot determine the tilts and relative heights of the two camera positions, so we have to assume as in radial triangulation that the photographs are true verticals and that the flying height was constant. With well-taken photographs we shall not be far out. However, we do not have to assume a constant ground height or photographic scale if we have either an accurate map or have made a radial line plot. In order to understand the measurement of relative height and to see what sort of errors may occur, we have to look more closely at the geometry and in particular to convert what we have so far called changes in the parallactic angle (or angle of squint) into small distances on the photographs because it is these that are actually measured with the parallax bar and which we then convert into height differences. In Chapter 3 we emphasized the need for anyone interested in surveying to appreciate at an early stage the relationship between angles and distances, and we find that the same is true in photogrammetry. For example, we talk of the eye measuring small angles but it does this by perceiving small distances between the images cast on the retina at the back of the eye by the light passing through the eye's lens on to different rods and cones on the retina. In large angles the brain records the angle through which the eye has rotated.

To find the true height of an object on the ground we have to carry out two separate operations:

USING STEREOSCOPIC PAIRS OF PHOTOGRAPHS   205

(i) find the height of the object at photographic or model scale;
(ii) convert the model height to a real height by dividing it by the photographic scale.

The first of these operations is the more complicated. Figure 8.3 shows what happens in real life when two vertical photographs are taken of a thin vertical object like a factory chimney or a tree lying under the flight line or base line joining the two PPs. The rays from the top and

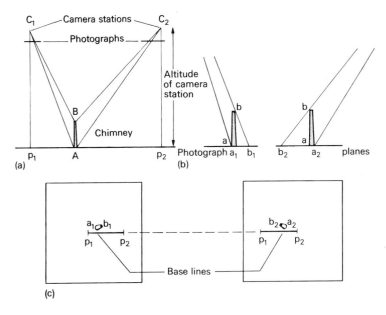

Fig. 8.3 Stereoscopic height appreciation on aerial photographs: (a) factory chimney as seen from the camera stations; (b) the model; a_1b_1 and a_2b_2 are images of the bottom and the top on the two photographs; (c) images on the two photographs.

the bottom of this reach the camera at the two aerial stations C_1 and C_2 whose ground or plumb positions are taken as P_1 and P_2. Since the camera is assumed to be vertical at both stations, the images of these plumb points on the photographs will coincide with their principal points P_1 and P_2 in the second diagram, where we look more closely at what happens in the model. Here the chimney AB in real life becomes the scaled-down object ab, and the images of it on the two photographs become a_1b_1 and a_2b_2; in the most common case these are such that the top b is displaced inwards and away from the PP on both photographs. It is clear that the triangles formed by each camera station, its ground position, and the object, and those formed by the top of the object and the images of its top and bottom on the photographs are similar in shape, and therefore

$$\frac{a_1b_1}{ab} = \frac{P_1A}{P_1C_1}$$

$$\frac{a_2b_2}{ab} = \frac{P_2A}{P_2C_2}$$

P_1C_1 and P_2C_2 can be taken as almost equal, and if we replace them by the mean height PC we can add the two equations together to obtain

$$\frac{a_1b_1 + a_2b_2}{ab} = \frac{P_1A + P_2A}{PC} = \frac{p_1p_2}{f} \text{ at photographic scale}$$

where p_1p_2 is the base length on the photographs and f is their focal length. Thus the total change of parallax between the top and bottom of the object on the two photographs gives us the model height of the object if we also measure the base lengths of the photographs and know their focal length. In the simplest case a is on or near the line joining P_1 and P_2, but normally of course it will not be in this position. When it is not we are only concerned with the movements of b in a direction parallel to the base line, because it is clear that this is what the parallax bar measures. This direction is defined by photogrammetrists as the *x* axis; the parallax parallel to it is called the *x parallax*, and the parallax at right angles to it along the *y* axis is called the *y parallax*.

Clearly we could measure the *x* components of both a_1b_1 and a_2b_2 with a magnifying glass and a fine scale, but this is more tedious and less accurate than viewing the photographs stereoscopically together with a floating dot (see p. 199) which can be made to touch the top and bottom of the chimney in turn. The relative movement between the two images of the dot required to do this is measured. The parallax bar does this in the same way that the floating dot is used in a precise stereoscopic plotting instrument and it measures the total change in the x component of the parallax $a_1b_1 + a_2b_2$ without measuring each one separately. Therefore all we have to do to obtain the model height of the chimney is to measure this total parallax difference and the base lengths of the photographs provided that we know their focal length. The parallax bar is shown in Fig. 8.4. It consists of two glass

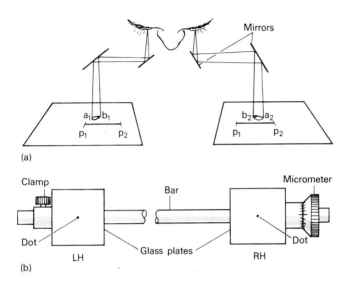

Fig. 8.4 Using the parallax bar. (a) Mirror stereoscope for viewing a and b stereoscopically. (b) The parallax bar: the left-hand (LH) plate is clamped to give the approximate distance; the micrometer is then used to measure the total difference x in the parallax between a and b, i.e. $a_1b_1 + a_2b_2$.

plates with a small dot, cross, or circle engraved on the underside of each and therefore in contact with the photograph. The plates slide along a bar, one being free to move over a few centimetres for coarse adjustment except when clamped, and the other permanently clamped but able to move over short distances for fine adjustment when the micrometer attached to it is rotated. The two glass plates are first set roughly over the same images on the two photographs which are set out with their base lines on the same line beneath a mirror stereoscope as described on p. 202, and the micrometer is then used to bring the two dots exactly together so that they are fused and appear to be in contact with the stereoscopic model. By placing the dot alternately on the top and bottom of the chimney and recording the differences between the micrometer readings, several determinations of the x parallax difference can be obtained more accurately and easily than in any other way (see also Plate 1).

It has been shown on p. 206 that to convert this parallax distance to the model height of the object we have to multiply it by the focal length of the camera and divide it by the base length. The focal length is normally recorded on the margin of the photographs and will usually be 6 in or 152 mm; if a normal-angle lens is used it will be 12 in or 304 mm, and if a super-wide-angle camera is used with a 3.5 in lens it will be 88 mm. The overlap of the photographs is normally 60 per cent leaving 40 per cent of their side length as the distance between each principal point and the images, on the same photograph, of the principal points of the photographs preceding and following it. With a 9 in square photograph 40 per cent of the side length is about 3.5 in or 90 mm, and this would normally be the distance between the images of the two principal points on both the photographs forming an overlap. A difference will be found in virtually every case owing to a combination of tilt (the photographs were not exactly vertical) and the differing heights of both the ground and the camera stations. This difference gives an idea of the probable errors in height determinations; normally it will only be a few per cent of the base lengths and a mean of the two can be used. Therefore the model height of an object will be about 5/3 of the parallax measurement with a wide-angle lens, over twice it with a normal-angle lens, and equal to it with a super-wide-angle lens. It is also evident from these ratios that, as is really obvious, the wider the angle of the lens the more accurate will be the determination of height for the same photographic scale and amount of overlap; and also, that if the amount of overlap is increased, thus shortening the base length, then the accuracy of height determination will be decreased. Thus scientists interested in local heights or slopes will find wide-angle or super-wide-angle lens photography the most accurate, and they should avoid using normal-angle lens photography if possible.

Having measured the model height of the object, the next step is to convert this to a real height by finding the model scale and dividing the model height by this. As mentioned in Chapter 7 there are several ways of doing this. The first, and least reliable, is to take the recorded height at which the photographs were taken from either the photographs themselves or as supplied with them by the contractor, and to subtract from this the height of the ground taken from the best available maps of the area. The ratio of this height above ground divided into the focal length (f/H) then gives the scale of the photographs. If good maps exist (say on 1/100 000 or 1/50 000 or a larger scale) it is more reliable to compare the photographs with these and obtain their scale directly. If sufficient identifiable detail exists then the comparison can be made actually in the area and at a level near the objects being heighted or slopes being measured, and this will be worth doing in mountainous country where there may be considerable scale changes across even a single photograph or overlap. If no such maps are available but a radial line plot has been or can be made and tied down to some sort of ground control, then comparing the distance between the PPs and minor control points in it as measured both on the photographs and on the plot will give the most reliable and accurate value for the scale of the overlap. Again, as with the base lengths, there will be small differences between the distances found from different pairs of points unless the photographs were perfect verticals and the ground flat, and the distances to use will be those between the minor control points closest to the object. In hilly country it may be necessary to intersect special detail points near the object in order to obtain the best possible value for the scale of the photographs in that part of the overlap. The photographic scale is the scale of the plot multiplied by the ratio of the distance on the photograph to the distance on the plot.

For simplicity we have so far considered an ideal tall thin object like a factory chimney or thin tree whose top is directly above its bottom such that both can be seen in the stereoscope, but in real life it may be necessary to compare the top of a tree or other object with a point on the ground somewhere near its bottom and apparently at the same level. So long as the 'bottom' point is not more than a few millimetres on the photographs away from the top, the ground is reasonably flat, and tilts of the photography are small, the errors introduced will be small, but of course it may often not be possible to estimate the size of the last two factors, and therefore the accuracy cannot be predicted. In general, however, the taller the object the more accurately in proportion can its height be measured. The accuracy of the parallax measurement is affected by the stereoscopic acuity of the observer, the magnification of the stereoscope, whether the photographs are viewed by reflected (as with prints) or transmitted (as with diapositives) light, and the texture of the images at the top and bottom points. No absolute figures can be given. In open forest the wider-angle lenses have the advantage when trees are scattered because there is more chance of seeing the base of the tree under its canopy, but in dense forest there is more chance of being able to see the same point on the ground on both photographs in a small clearing with a narrower-angle lens. In general there is little advantage in using photographs other than those taken with an ordinary wide-angle (6 in) lens.

9. THREE-DIMENSIONAL MAPPING

The plotting instruments

Although in most cases the actual use of three-dimensional plotting instruments is outside the province of field scientists, some larger university departments, particularly in physical geography and forestry, do include air survey sections which use them. Specialist technicians are normally employed to operate the instruments; but it is a poor professional scientist who does not have at least some understanding of how the measurements on which he relies were obtained and how accurate they are. Field scientists should also be aware of what is possible as some sort of check on the claims made by organizations preparing maps or making special measurements for them. In Chapter 8 we saw in outline how such instruments operate, and in this chapter we shall look in more detail at the five principal types and particularly at those which are easiest to understand and use and therefore the ones most suitable for someone not trained in photogrammetry to operate himself. For detailed instructions on how to set them up and operate them the reader is referred to specialist textbooks, but in the next section we shall give some idea of how they are operated and in particular how this affects the sort of ground control which a non-surveyor who has absorbed the earlier chapters might be expected to establish. Where a detailed survey of some particular feature is required, and where aerial photography already exists, this may be far more economical of his time than carrying out a detailed field survey using a plane table or chain, compass, and level. The theory of three-dimensional photogrammetry was outlined in pp. 189-93. How in fact is this translated in practice, and what instruments are used?

There are five main types of instrument. The first, the *projection* instruments, although now almost obsolete, are by far the easiest to understand and operate and have the advantage of creating a visible model; and several teaching departments have instruments discarded by mapping organizations. Next to these in ease of understanding are the *mechanical* and *optical–mechanical* or *analogue* instruments in which, as in the projection type, the glass diapositives of a pair of photographs are set up in carriers which physically reproduce the

212   THREE-DIMENSIONAL MAPPING

exact scaled-down relative positions and attitudes of the camera at the original two camera stations, and space rods or rays of light represent lines of sight to points in the model. In the other two types, the *approximate* and the *analytical* instruments, this is not done, and the photographs are set on parallel tables which move only in their own plane; the effects of the small positional and attitudinal differences from the ideal set-up are taken care of either by complicated (but relatively easily manufactured) mechanical devices in the first, or by very sophisticated processes of electronic digital recording and simultaneous (*real-time*) computation in a relatively large computer. Thus instruments of the first type are fairly cheap and approximate (and often use paper prints), while those of the second type are capable of a very high accuracy but are very expensive and difficult to operate correctly.

The basic principle of projection-type plotters is delightfully simple which is why they are still used for teaching and are described in detail here. Glass diapositives are placed in a pair of slide projectors whose lenses have the same geometrical characteristics as those of the original cameras, though in the earliest case of the Multiplex (Fig. 9.1) their

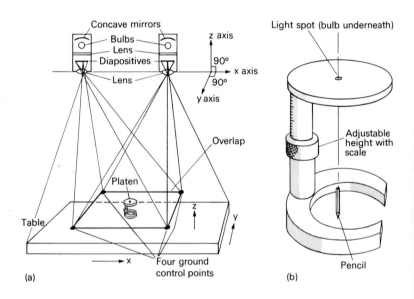

Fig. 9.1 The Multiplex: (a) geometry of the Multiplex; (b) details of the platen.

focal length was reduced to nearly a quarter of that of the original camera and the diapositives were reduced in size in a similar ratio from the original 9 in × 9 in photographs to $2\frac{1}{2}$ in × $2\frac{1}{2}$ in. The two projectors are set up on a bar above a flat horizontal table so that their effective heights above this and their distance apart are scaled-down versions of the original altitudes and separation of the two camera stations. By means which cannot be described in detail, the original small *tilts* and *tips* (the former at right angles to the flight line and the latter along it) of the cameras at the two moments of exposure are exactly reproduced, as is any change in azimuth or crab and the difference between their heights above sea level. Each diapositive is then projected down onto the table by a light above it, one through a red and the other through a blue-green filter (or by using complementary Polaroid filters), so that the two images on the table are in complementary colours.† In fact the projectors are not at the correct height above the table itself but are a few inches above this so that rays of light to the same part of the ground from the corresponding images on the two diapositives coincide, not on the surface itself, but on a small platen (Fig. 9.1(b)) standing on the table and capable of being moved to any point on it. This platen has a fine pinhole in its centre lit from underneath which forms the floating spot; it also has a micrometer adjustment for height and an altitude scale, and there is a pencil point directly under the light spot. When the height scale reads zero the surface of the platen defines the horizontal datum (usually sea level) above which heights on the model will be recorded, and it is on the platen surface that the two coloured overlapping images are viewed. When the platen is adjusted so that it is touching the surface of the model, a single combined grey image is seen with a single naked eye, but to increase acuity and create a three-dimensional model the operator wears a pair of spectacles with red and blue-green filters so that each eye sees only one picture as in a normal stereoscope. A more expensive and cumbrous alternative which can be used with coloured photographs is to use a synchronizing flicker mechanism for both the projectors and the spectacles, but the colour filter system is the one most commonly used.

To trace a feature on the ground the spot is moved along it and kept in contact with the model by adjusting the height scale as required;

† A similar use of complementary colours – the *anaglyph* system – is employed when stereo pairs of photographs are published in a book and coloured viewers are provided with the book.

spot heights can be read off this if required. To draw contours the height scale is set at the equivalent height (allowing for the scale of the model) and left fixed while the platen is moved about so that the spot touches the model all the time and the pencil beneath it traces out the contour. An advantage of the smaller size of the diapositives and the shorter focal length of the projectors is the greater depth of focus since this makes it possible to map terrain with considerable height differences; however, the resolution and viewing conditions are relatively poor so that the Multiplex is definitely a third-order instrument. In all of this of course we assume that the model is correctly set up; this is a complicated procedure and all we need to realize here is that it is done by knowing the positions and heights of four points — similar to the corner pass points of radial triangulation — in the corners of the overlap and identifying them on the photographs, and then bringing the model down onto them by checking each with the platen and spot so that it fits exactly. Once this is done correctly every feature on the model is at its correct place and height, and all the operator has to do to map it is to keep the spot touching them and let the pencil trace the features to be mapped on the table below. The resulting map will be approximately $2\frac{1}{2}$ times the size of the overlap and so will cover an area of about 17 in × 8 in (40 cm × 20 cm). The main use of the Multiplex instrument is for medium- or small-scale mapping, and it is very suitable for mapping forests since many of these are in mountainous areas and when only the tree canopy is visible no very great contouring accuracy is possible anyway. However, the Multiplex is not suitable for more precise measurement of individual tree heights because of its poor resolution; and a parallax bar used with a powerful stereoscope would produce better results.

An important advantage of the small size of the Multiplex projectors is that several can be set up on one long bar and used to run what amounts to a three-dimensional traverse between two parts of the ground control framework, thus making it possible to space this more widely than would otherwise be necessary (Fig. 9.2). This corresponds, of course, to 'bridging' or aerotriangulation by slotted templates between single ground control points as described in Chapter 7 and also to the running of a traverse or triangulation chain on the ground between two framework points of a higher order. After having positioned, scaled, and levelled one overlap onto four ground control points at the corners of the model it is clear that we can set up a third

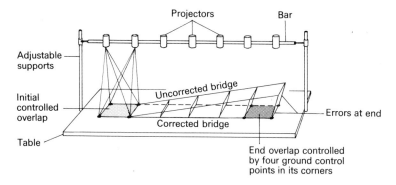

Fig. 9.2 Multiplex bar bridge.

projector with the third photograph to form a new model with the second photograph of the controlled overlap, and this model will also be positioned, scaled, and levelled, although with inevitable small errors. A third model can then be fitted in the same way by setting up a fourth projector with the fourth photograph in the strip and fitting this to the third one, again of course introducing small errors of scale, swing, and dislevellment (see Fig. 9.2). Up to seven or eight successive projectors can be set up on the bar in this way until the last two include an overlap which has also been controlled by four ground-surveyed corner points. When the four points on this model are compared with the ground survey co-ordinates, errors will be found in position, scale, and height, and these will be noted and corrected. The corrections applied to bring the model down on to them will then be distributed back through the models in between by adjusting each projector in proportion, in exactly the same way (except that it is three-dimensional) as is done with the intervening templates of a slotted template strip, or as is done with the computed co-ordinates or plotted positions of the stations of a ground survey traverse. This is sometimes known as *bridging* or aerotriangulation; it is done entirely graphically with the Multiplex and requires no computation except the distribution of the corrections to be made to the various scale readings on each projector. This can be done mathematically now that high-speed computers exist which can handle the complex mathematics involved in measuring recorded attitudes and positions of successive models. However, for non-mathematicians the graphical bridging technique of the Multiplex will add to its attractions and some secondhand instruments may still be available.

The *Balplex* projection instrument (Fig. 9.3(a)) is an improved and more expensive type of Multiplex with increased brightness achieved by a sophisticated ellipsoidal mirror in each projector. However, to achieve greater accuracy both in large-scale planimetry and in contouring, a different projection system must be used – the *Kelsh plotter* (Fig. 9.3(b)). In this instrument full-size diapositives of the original negatives can be used because only a part of the overlap is illuminated. Narrow-beam projectors are aligned by rods joining them to the platen so that – like the two spots following an actor or ballerina – only the area of interest is illuminated, but apart from this the principle is the same. With the Kelsh plotter the model scale is about six times that of the photographs, and whereas with the Multiplex heights can only be measured to somewhere below one-thousandth of the flying height, with the Kelsh plotter this accuracy can be doubled. However, the Kelsh plotter cannot be used for bridging in the simple way described on p. 214.

The advantages of all projection instruments are their simplicity of theory and operation, their lack of precisely shaped moving parts which can wear or warp and cause errors, and their consequent comparative cheapness. Their disadvantages are that the anaglyph system cannot be used with colour photographs, the plotting scale is fixed

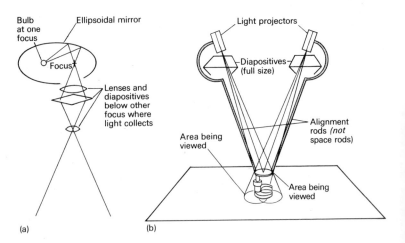

Fig. 9.3 (a) the Balplex projector; (b) the Kelsh plotter.

within narrow limits relative to the photographic scale, they have poor viewing characteristics, they have a limited height range when full-size diapositives and projectors are used, they have to be operated in semi-darkness requiring separate booths which make supervision more difficult, and there is a continual noise from the blowers required to keep the projectors and the slides at reasonable temperatures.

The more accurate analogue or mechanical instruments do not have these disadvantages but they are more expensive, require air conditioning to keep dust out of their moving parts, and are more difficult to operate, maintain, and understand. Until the advent of the analytical types they and the optical/mechanical models were the most accurate of all plotting instruments in their most elaborate and expensive first-order versions, which cost tens of thousands of pounds. They are surveying equipment, and except perhaps in the lower-order and cheaper versions are not really suitable for non-surveyors. Instead of diapositive projectors representing the cameras, the mechanical instruments have a mechanical equivalent or *analogue*. The diapositives, which are usually full size, are placed in replicas of the original cameras but with each lens replaced by a universal joint in which slides a *space rod*; the upper end of this is directly below the point on each photograph which is being viewed (Fig. 9.4 and Plate 2), while the two lower ends pivot in a common joint so that the intersection of their axes represents the three-dimensional position of the point in the model beneath the two diapositive holders. In some versions a parallel mechanism is used between the space rods and these holders, but the principle is the same. This is the basic principle which is easy to follow and, while the model is not actually visible, the intersection of the space rods can be seen to trace the surface of the model as the operator scans different parts of it. In more elaborate versions this intersection is moved along the flight line direction (the x axis) by one hand wheel, and at right angles in the horizontal plane (the y axis) by another; a foot wheel is used to change its altitude (along the z axis) see Plate 4. By means which are too complicated to describe here (but which in essence resemble that of the parallax bar described on p. 206) the two halves of the floating spot are incorporated in the separate optical systems through which each eye views the corresponding photograph, and the foot wheel enables the operator to raise or lower this onto the model surface in exactly the same way as is done on the parallax bar by turning the

218 THREE-DIMENSIONAL MAPPING

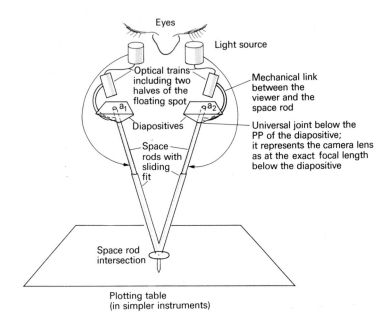

Fig. 9.4 Optical–mechanical plotting instruments.

micrometer. Thus the operator sees, at magnifications which can be adjusted up to about 10 ×, a small circular section of the whole overlap with a single spot in the centre, and within this circle he can see the shape of the ground and of any features on it, and adjust the spot to follow these exactly.

In some of the cheaper forms of these instruments (like the Wild B8 or Kern PG 2) the space rod intersection is moved directly by hand on a table, and a pencil beneath it or connected by a pantograph draws the map as in the projection instruments. However, in higher-order instruments hand wheels are used to move the space-rod intersection and are directly geared to, or operate by servomotors, a plotting table at the side of the instrument on which the map is drawn. This is the most accurate way of transferring movements from the model to the map, and it allows for greater degress of enlargement or reduction than the pantograph. It is also possible to link the gears and hand wheels to *digital encoders* so that the instrument can be used to record co-ordinates instead of, or at the same time as, drawing the map or plotting points graphically on it. Thus, as indicated on p. 215, starting from one overlap which has been positioned, scaled, and heighted by ground control points, it can build a *bridge* of co-ordinated points on subsequent overlaps until another controlled overlap is reached. Alternatively, all the overlaps can be tied together at a common scale and then mathematically made to fit on to a number of scattered control points. This is often one of the main tasks of the most accurate first-order instruments; the plotting of individual overlaps based on minor control produced by these is then done in detail by less accurate and less expensive second- or third-order instruments.

The *optical-mechanical* instruments are less common than the purely mechanical ones and less easy to understand, since although there are space rods or links these do not directly represent the rays of light directed from the corresponding images on the two diapositives towards the model position of the point being viewed stereoscopically. This function is taken over by rays of light projected through lenses which are duplicates of the original camera lenses, and then through a complicated optical system so that the model is not only invisible (as it is in the projection instruments) but it cannot even be seen to be traced out by the movements of the intersection of the space rods. Apart from these fundamental differences in design, for the layman there is no obvious difference from the more elaborate purely mechanical instruments; both types have hand wheels and a foot wheel for horizontal scanning with the floating spot and for moving this in and out of the model, and both drive the separate plotting table on which the map is drawn through gears or more elaborate mechanisms.

Clearly all these instruments require a high degree of skill both in operation and in maintenance, and only the simpler forms are suitable for the small photogrammetric section inside a non-survey organization. However, the non-surveying scientist should be aware of the tremendous potential of such instruments for they can map accurately up to ten times the photographic scale and in ideal conditions plot accurate contours at an interval something like 1/2000 of the flying height above the ground. This means that a contractor with one of these instruments, given the right ground control, can produce a map from 1/40 000 scale photography (which is in use in many countries) at 1/4000 scale with 10 ft (3 m) contours covering an area of 4 miles × 2 miles (6 km × 3 km) on an area of 60 in × 30 in (nearly 2 m × 1 m) from a single overlap for which only four positioned and heighted ground control points are required. This could obviously save a great deal of detailed mapping on the ground of such features as dunes, glaciers, or human settlements. Moreover, if previous photography exists, changes of volume and height as well as of position can be shown with precision. These precise instruments can also be used, as mentioned also on p. 188, to plot on a field sheet small features such as termite mounds or bushes which the field scientist can then use as a framework to control detail mapping of features (such as soil boundaries or pits) not identifiable on the photographs, without all the work required to put in his own framework on the ground.

Little need be said about *analytical* plotters (Fig. 9.5). These measure only *x* and *y* co-ordinates on each diapositive and then transform them into the three-dimensional co-ordinates of each common point in the model formed by the two photographs by means which are far too complicated to describe here. They have the advantage that it is relatively easy to make the original co-ordinate measurements very precise and that all subsequent operations, being mathematical, can in theory be made free from errors, unlike the analogue mechanical conversions which are subject to wear and warping in joints and space rods. Although

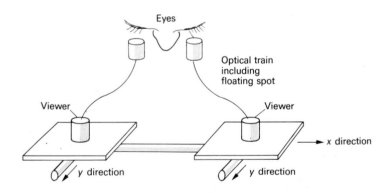

Fig. 9.5 Geometry of analytical and approximate instruments. Both diapositives (or paper prints) are in the same plane and can only move in the *x* and *y* directions together and relative to each other. All the rest is taken care of by mechanical or analytical methods too complicated to illustrate here.

the cost of precise engineering of this type continues to increase while that of digital recording devices and computers is decreasing, analytical plotters are still outside the budgets even of quite large survey firms, and it seems likely that they, and the skilled and well-educated operators required to obtain the best from them, will be limited to organizations engaged in topographical mapping or special engineering measurement. It is not possible at this stage to say whether the fact that a contractor is using such a plotter is an advantage.

Approximate plotters (Fig. 9.5) are relatively cheap and several useful models exist. They employ various theoretically elaborate although physically simple techniques to arrive at an approximate three-dimensional solution. Any scientist thinking of using or buying one should study the papers written by their designers and consult his survey and photogrammetric colleagues because some of these instruments have practical disadvantages well known to those who have operated them but which are not widely publicized. Compared with the Multiplex, approximate plotters can be used in normal office conditions and take paper prints; however they do not produce a visible model and are not easy to understand and can therefore more easily be used incorrectly. In some cases the optics are inferior to those of higher-order instruments, and this may be important where small-scale photography has to be used for mapping features not easily seen on the photographs.

Rectified photographs, orthophotos and mosaics

So far consideration has only been given to *line mapping*, but this is not always the most suitable product and it is almost invariably slower to produce than *photo-maps* made more directly from the air photographs themselves. Photo-map is a general term meaning a map in which the photographs are reproduced directly with names and some symbols added; the accuracy varies according to the method of compilation. In line maps the interpretation has been carried out by a cartographer (if not by the scientist himself) without specialized knowledge, but the field scientist may feel that the photograph, from which nothing has been removed, may suit his purposes better, especially if he is accustomed to handling and interpreting air photographs. It will contain more information, but as we have seen it is almost invariably distorted so that scale and shapes vary and it is not a proper map from which accurate measurements of angle, distance, or area can be made. A photo-map is a compilation of the photographs in which most of these distortions have been removed. There are two possible ways of removing the distortions from an ordinary near-vertical air photograph:

(i) *Rectification*, in which the negative is projected through a *rectifier* (a special form of enlarger) on to bromide paper to produce the equivalent photograph as if it had been taken with the camera axis truly vertical. It is still a perspective view from one point and therefore the usual height distortions will be present (see Plate 3).

(ii) *The orthophotograph*, in which the original photograph is projected successively in a series of small sections so that the scale everywhere is the same and it is in effect projected orthogonally like a map; apart from minor distortions inside each section it will be true to scale and therefore exactly like a map, except that it shows every feature appearing on the original photograph.

Rectification is more suitable when the ground is relatively flat, because it is usually much easier and cheaper to do this than to produce an orthophotograph. The negative is inserted into the rectifier and projected down onto a scaled plot on which a minimum of four identified minor (or ground) control points in the corners of the photograph have been plotted; normally they would have been taken from a radial line plot, but they could be ground control points or even detail from an existing map at the required scale if this is available. Once an accurate fit has been obtained to these a sheet of photographic paper or plastic is placed on the table and the negative is exposed on to it, producing a rectified enlargement. Certain conditions have to be fulfilled in the rectifier in order to achieve accuracy, so this is not just an enlarger with a tilting table; both the negative and the printing table must rotate relative to the projecting lens and thus the equipment is quite expensive. Most air survey firms possess a rectifier, and since it is relatively easy to operate, an agricultural department with a photographic/photogrammetric section might well think one worth having, especially if many of the projects in hand cover relatively flat ground as on river flood plains or irrigation schemes. The only control required is that provided by a slotted template lay-down (which of course provides nine points per photograph), and although in theory it is wrong to enlarge a control plot and use it as the basis for a map, in fact since the photographic images are generally less precise than line maps a careful enlargement by two or three times of the lay-down could be used to produce a single satisfactory rectified or controlled photograph or a mosaic formed from several of them. An alternative is to use enlargements and enlarged templates of the original photographs for the lay-down itself. No height control is required since the effect of small differences in the height of the minor control can be ignored over a flat area.

Where the ground is not flat a more elaborate technique has to be used involving, as with accurate contour mapping, a three-dimensionsal solution based on heighted ground control. A three-dimensional model is produced in the way already described, and small sections of this are photographed successively at the correct scale of the map by varying the degree of enlargement. Various techniques are used depending on the type of plotting instrument providing the model, since most orthophotoscopes are designed round an existing plotting instrument such as a Kelsh or an analogue optical–mechanical type. In most of them the model is traversed by a slit which is scanned transversely so as to divide the whole overlap into small rectangles; each of these is projected onto a positive in turn when it is touching the model, and so the whole positive of the overlap is built up until it is complete. Here again it should be noted that whereas in rectification of single photographs the unit is the photograph, here it is the overlap. In a more elaborate and sophisticated technique called the Gestalt process, the model is divided up into a series of hexagons like a honeycomb; scale differences within each of these are allowed for by a system of fibre optics so that the smallest unit is less than a square millimetre at photographic scale. Which technique has been used can usually be detected by careful inspection of the orthophotograph, when either the edges of the successive strips or of the hexagons can usually be seen.

In either case, whether the original photograph has been rectified or the overlap has been projected through an orthophotoscope, the result appears to be like an ordinary enlargement unless checked under a magnifier; the orthophotograph is of course true to scale everywhere like an accurate line map but the rectified photograph will only be so if the ground is flat. The ordinary map user not accustomed to air photographs may find that the images are unfamiliar and he will have to do his own interpretation instead of having it done for him by the cartographer using various lines and symbols, but of course this is just the advantage sought by the specialist field scientist who wishes to study the original imagery and not be presented with an interpretation and selection in which the first may be incorrect and the second leave out the features in which he is most interested. Clearly, even the orthophotograph cannot resolve areas where there is a sudden change of scale such as steep cliffs or the edge of a tall building with vertical sides where the change is instantaneous. Only a cartographer can resolve this kind of problem, and even orthophotographs are much less useful in areas of high relief such as mountains or city centres with tall buildings. A further point is that since first- or second-order three-dimensional plotting machines are used for producing orthophotographs there is a potential for considerable enlargement, and because it takes some time to set up each overlap in such instruments it is very uneconomic not to use this facility rather than to produce a result near the scale of the original photographs.

These considerations make clear the conditions under which rectified enlargements or orthophotographs should be used. First the user should be able to interpret air photographs (and he can achieve stereoscopic viewing by using individual prints or enlargements at the same scale in conjunction with each rectified enlargement or orthophotograph), secondly he should require to interpret features (e.g. soil, vegetation, building types, etc.) which the ordinary cartographer would be unable to distinguish reliably, and thirdly he should require a larger scale than that of the original photography. Fourthly the ground must not be too mountainous or heavily built up, or even the orthophotographs will show too many inconsistencies at the margins of the rectangles or hexagons from which they are compiled; these normally show as the presence of jagged edges on a line which ought to be straight such as a steep section of road or the edge of a flat-roofed building. In fact the ideal terrain, and one in which orthophotographs have been used a good deal, is in undulating areas farmed by smallholders in an uneconomic and often fragmented way in which better agricultural practices are being proposed. The better survey firms will give an honest opinion on whether simple rectified enlargements based on a radial line plot and relatively sparse ground control would serve in a given area, or whether the considerable additional expense of denser heighted ground control, aerotriangulation, and the use of a three-dimensional orthophotoscope are justified. The scientist must have a clear idea of just how accurate he really needs the result to be and if possible commission beforehand a few samples of both techniques over the area for testing. After all, any measurements he may be taking in the field are likely to be with relatively simple and rough techniques such as chaining or pacing and a hand compass, which are accurate only to about 1 part in 50 or 1 part in 100 over small areas, so there may not be much point in having the photo-maps much more accurate locally than this. It is worth remembering also that since the distortions in rectified photographs over ground which is not quite flat are related to the ratio of the height differences to the height from which the photographs were taken, the smaller the original scale (or the longer the focal length) and the greater the degree of enlargement used to produce the rectified enlargements, the more accurate these will be – and of course the fewer ground control and minor control points will be required.

As in the case of ordinary planimetric line maps, contours can be added to orthophotographs produced three dimensionally, and in many cases this is done as a separate operation though often without having to set the photographs up in the machine a second time. However, a cheaper solution is to draw *dropline contours* as the scanning of the overlap proceeds. These are profiles drawn along each position of the slit as it is scanned, with the sections between successive contours shown by different widths of line. When the overlap is complete the draughtsman joins up the ends of each section of the same width and these lines form the contours. In general it is preferable to follow the contour line itself in the instrument, particularly if detailed shapes of topographic features are important, since these will not appear in the dropline system.

In summary, both rectified enlargements and orthophotographs can be a very useful alternatives to line mapping, but before entering into a contract the scientist should consider very carefully exactly what he wants and when. In general all kinds of photo-maps can be produced more quickly than fair-drawn line maps because considerably less time is required than for fair drawing which is a slow manual process.

Mosaics

A *mosaic* is an assembly of air photographs cut up and stuck together so as to form a continuous picture of the ground. The cutting is done partially and the photograph is then torn along the cut so as to produce a feathered edge. In the early stages of a project covering poorly mapped country, when no ground control is available, the best that can be done is to fit the photographs together by eye, cutting and joining them where they seem to fit best and where there is relatively little detail prominent enough to show up the inevitable inconsistencies along the joins. Only if the photographs were truly vertical and the ground completely flat would the edges fit perfectly. This is an *uncontrolled* mosaic and it can be a very useful way of producing some sort of map soon after the photographs have been taken and for planning and co-ordinating the activities of a field party in an area where only small-scale maps exist. If the area is large the mosaic can be divided up into separate sheets and re-photographed to produce a *photo-map*, but it will not be an accurate one. Where the individual photographs have been rectified they can be assembled into a *controlled mosaic* where the fit will be much better, especially if the ground is reasonably flat, and the result can be a photo-map accurate enough for many purposes. However, in hilly or mountainous country, or where a really accurate photo-map is essential (e.g. for measuring areas), then it must be compiled from orthophotographs. If existing medium-scale maps are good enough (e.g. good 1/50 000 maps) and a photo-map at (say) 1/10 000 is required for an agricultural project, careful planning of the photography could result in each orthophotograph covering a sufficient area to form a single map sheet, since an enlargement of five times is quite acceptable from good photography. If only a few copies are required they should be printed photographically on bromide paper, but if a large number are required (and are not to be used with magnification or a stereoscope) then half-tone printing (see p. 301) will be very much cheaper although the result will be inferior. The difference between a mosaic and a photo-map is small, but in a photo-map the edges of individual photographs should be almost indistinguishable and names, roads, etc. are usually added by hand.

Ground control for and from air photographs

Although only a general description of how three-dimensional air surveys are carried out has been given in this book, in this section we explain in detail how scientists can provide the necessary ground control for them in cases where the cost of sending out a land surveyor for a comparatively small task overseas is not justified. The way in which the structure of a slotted template assembly decides the layout of the ground control required to scale and position it was described on pp. 184-7, and in the same way the structure of three-dimensional photogrammetry defines its ground control requirements. The individual item here is not the single photograph but the overlap, and control for this is a minimum of three, and preferably four, points. Where the scale is near that of the photography and/or the contour interval is large, then overlap control can be spaced widely and the photogrammetrist will 'bridge' between these in the way already indicated on p. 214. In large areas this will mean a fairly extensive network of ground control, and in this case the whole project will be too large for the non-surveyor and should be handed over completely to a professional survey organization; however, the actual requirements for each control point are very much the same as in small schemes.

From this it is clear that the principal operation of interest to the field scientist is where one or two overlaps only are to be controlled by him so that a particular area of ground can be mapped in detail at a much larger scale than that of the photographs, and only half a dozen or so control points are required. In this case (as already described on p. 214) the photogrammetrist will set up the first overlap in his instrument and bring the two photographs into correspondence so that he achieves a perfect model all over the common overlap. This process is termed *relative orientation*. Having done this, he then compares the model with each of the control points in turn and adjusts its position, scale, and *absolute orientation* until it fits these exactly in both position and height. Once this is done he can draw any feature by simply following it with the floating spot, and draw contours by following the surface of the model while keeping the height of the spot fixed. What then are his requirements for ground control in location, accuracy, and identification?

In *location* the points must be in the corners of the overlap, outside the area to be mapped. Where more than one overlap is to be used the photogrammetrist will choose points in the supralap along one strip, and in the lateral overlap between adjoining strips. Unless the photography has been very carefully planned and taken using accurate existing maps, the overlaps are likely to be out of phase along adjoining strips and it will not be possible to use all the points in the initial strip, but they will of course be used as a further check. If very accurate heighting is required the photogrammetrist will need a fifth point somewhere in the middle of the overlap, but although this must of course be identified on the photographs its position does not have to be fixed on the ground since only its height will be used, and its position will be fixed with sufficient accuracy by the photogrammetrist once he has put the overlap down onto the four corner points.

The *accuracy* requirement will depend on the degree of enlargement and the proposed contour interval, and on the type of instrument being used. In the extreme case where enlargement up to eight times is planned, a single overlap whose real size is about 7 in × 3.5 in (18 cm × 9 cm) will produce a map of dimensions 140 cm × 70 cm (1 m^2), and if accuracy over this is to be held to 0.5 mm plotting will have to be accurate to about 1 part in 3000. Control should always be at least twice as accurate as the process which it is controlling, and so we are talking about accuracies of the order of 1 part in 6000 across the distance between the two furthest points. This means that techniques such as plane-tabling or compass surveys are not accurate enough, and normally a theodolite would have to be used, unless the area is very flat when a careful chain (tape) survey could, though rather laboriously, achieve this accuracy. If the area extends over several overlaps then a correspondingly higher accuracy should be aimed at in theory, but in practical terms something of the order of one part in ten thousand (which can be achieved with a steel tape and a one second theodolite) would be adequate. The photogrammetrist is more interested in being able to obtain a good fit at the corners of each overlap rather than in what may happen between the outer limits of the whole survey. In height the most accurate plotting instruments can measure to about 1 part in 4000 of the flying height and contour about half as accurately, so the height requirement depends on this and on the contour interval rather than on the degree of planimetric enlargement. As a general rule ground control heights should be accurate to about a tenth of the proposed contour interval, or to a quarter of the proposed accuracy of spot heights, which cannot be much better than the limits given above. For 1/40 000 scale photography taken from 20 000 ft control should therefore be accurate to about 2 ft (0.6 m).

However accurate the fixing and heighting of the control on the ground, its value to the photogrammetrist must depend on its being unambiguously and precisely *identified* on the photographs, and this is a process whose importance is often underestimated. It goes without question that it must be done by stereoscopic viewing of pairs of photographs and not from single ones. There are really three different parts of this process: avoiding gross misidentification, selecting and describing in sufficient detail a suitable point for plan control, and selecting and describing a suitable point for height control. Gross misidentification — the wrong building, the wrong hill top, or the wrong tree or bush — is only avoided by taking great care. The identifier should always know where he is, he should work from the known and absolutely sure to the less sure, and he should be saying to himself not 'I seem to be here because this is what I see on the ground and it seems to agree with the photograph', but 'I know I have arrived here, and from the photograph this is what I should see around me — and I do.' In a developed area with recent large-scale photography this is relatively easy, but these ideal conditions are rare; one is more likely to be in an area where the only features are grass, rocks, trees, and bushes, with perhaps a few paths, and the photography may be on quite a small scale (1/40 000 or so) and several years old. To follow the first two principles the identifier must start from some absolutely certain point, even if it is a mile or more from the area where the control point is required, usually on a road or railway and at a turning or where there is a bridge or other building identifiable on the photographs. He should then 'navigate' on the photographs using a compass and either calibrated pacing (i.e. checked against a taped distance) or, on smaller-scale photographs, a vehicle with a speedometer reading to tenths of a mile or (better) of a kilometre. He should not proceed directly to the proposed point unless it is close by, but via a number of unambiguously identifiable objects — such as large trees or path intersections — which he can check when he arrives at each in turn on the principle of 'I am now here; do I see round me what I should?' Usually the proposed control point area will have been selected by the photogrammetrist both to suit himself and because from his experienced inspection of the photographs it seems to be an area where the ground features should be identifiable without too much difficulty. However, if, as quite often happens, they have changed too much since the photographs were taken — for example by clearing for crops — the identifier must pick another area as close as possible and select a point there.

The requirements for plan and height control are different, and where both are required two different points in the same area may have to be chosen. For plan the photogrammetrist likes ideally a symmetrical light or dark coloured feature a little larger than his dot, which is usually about 50 or 100 μm (i.e. 0.05 or 0.1 mm) in diameter at the scale of the photograph, so a bush, a tree crown, a boulder, a path intersection or a small clearing, or on large-scale photography even a cowpat or clump of weeds, should serve. Other alternatives (which would be easy to fix by a ground survey intersection) are the spire of a church or the minaret of a mosque, or a sharp pinnacle of rock on a hill top. Corners of houses or other structures are less precise because they are not symmetrical (and a white corner may *halate* or expand on the photograph), but they may be acceptable because to make up for this they are less ambiguous than a bush or tree. In difficult circumstances where mistakes are likely the identifier should follow the example of the professional land surveyors and identify a group of three points in the area required; this not only increases the chances of at least two being correct but also helps the confidence of the photogrammetrist. However carefully the survey and identification is done there are bound to be some mistakes (or there may be something wrong with the photography), and if all three separate points agree at each location then this is one cause which the photogrammetrist can eliminate in trying to find out why the overlap does not fit the control.

Many plan control points are also suitable for height, provided that they lie in an area of flat ground with sufficient texture for the photogrammetrist to be able to make a good contact with his floating spot; clearly the spire/pinnacle type of point is not suitable and a separate height point must be found nearby either on the ground or on a flat roof or similar structure. The precision with which this must be identified depends on the flatness of the ground, but there must of course be no gross misidentification. In general, height control points are much easier to identify with sufficient accuracy because nothing like the same horizontal precision is required. For both plan and height control points the field surveyor must produce for the photogrammetrist two documents:

(i) a marked photograph;
(ii) a diagram (Fig. 9.6(a)).

Different mapping organizations have different requirements, but in general they prefer for the first a very fine needle point pricked through the identified point on *one* photograph only (although of course the surveyor, particularly if overseas, will prick both of the pair and keep one as a precaution against the loss of the first). This fine hole should be circled on the *back* of the photograph and labelled with the appropriate number or letter referring to the point whose co-ordinates and/or height will be supplied on the list of ground control. A good mapping organization will provide standard forms for the second

THREE-DIMENSIONAL MAPPING 231

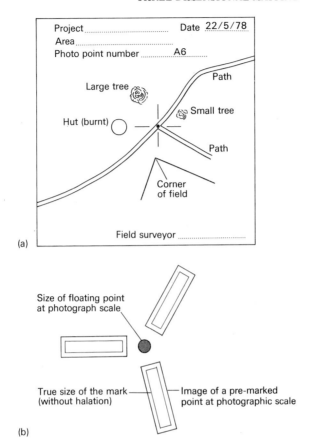

Fig. 9.6 Identification and premarking of photographs: (a) identification diagram for ground control; (b) premarked point.

document, usually with a carbon copying system. The main principle to be followed is that the diagram should show, at the sort of scale (6–10 times that of the photograph) which the photogrammetrist will be using when viewing the photographs, the features round the point which will help to identify it. These should be drawn by the surveyor while looking at the stereoscopic pair and confirming their existence on the ground. In older photography they may include features no longer visible whose previous existence can be proved by, for example, the stump of a tree which has been cut down or the floor of a grass hut that has been burnt or removed, but they must *not* include any features that the surveyor sees on the ground which do not appear on the photographs, for clearly these are of no interest to the photogrammetrist and will only confuse him. The object of the diagram is really to establish his confidence in the marked photograph by showing that the surveyor was really there and did see on the ground round the point all the features which the photogrammetrist is now seeing in his field of view as he looks at the area in his instrument and compares it with the marked-up photograph. A sample diagram is shown in Fig. 9.6(a).

It is clear from what has been said above that since large-scale plotting from high-order machines, where one overlap when mapped may cover a square metre or so, is accurate to something like 1 part in 3000, this is the equivalent of ground control accuracy higher than that of the plane table or compass and tape techniques, and we might reverse the usual procedure and use photogrammetry to control field surveys. There may well be circumstances in which this accuracy can be used and thus save the field scientist the labour and knowledge required to establish such a framework to control his own relatively low-order field measurements using simple techniques. There are two possibilities: establishing points marked on the ground before photography, or the more common circumstances in which photography already exists and sufficiently small and precise features can be found on it, be fixed by the photogrammetrist, and then used by the field surveyor to control his own surveys. This will be particularly useful in cases where the features he is mapping do not appear on the photographs, e.g. soil pits or other sample sites which he is establishing, or the exact boundary between different soils or vegetation which can only be established precisely on the ground, or lines of levels. The sort of features to be chosen for the photogrammetric framework will be the same as those used for photographic control points, except that all manmade features will normally be shown anyway, and the sort of features involved in this case are therefore those which the photogrammetrist would not show normally such as isolated bushes or small trees, individual fence posts in a straight fence, termite mounds, circular threshing floors, and so on. The field scientist must mark up on contact prints or enlargements those features that he wants fixed so that the photogrammetrist will select these and plot them accurately on the map. They can then be marked for field use by posts or flags and used by the field scientist for locating himself nearby using a compass, tape, or even pacing. This could be particularly useful and time-saving in an area which is relatively featureless from a topographic mapping viewpoint, such as a river flood plain or semi-desert, or alternatively in forested, marshy, or rugged areas where accurate field survey is difficult and laborious.

In the less common case where photographs are being taken after the scientist has been in the field, or while he is actually there, *premarks* or signalized points can be used for controlling both the photogrammetry and the field survey. These are precise and unambiguous air-visible marks put on the ground before (preferably just before) photography, and they can save a great deal of time both for the field man and for the photogrammetrist in identifying and in understanding the identifications and in eliminating all the errors which do occur. The best premark for precise control points is a Y with the arms not joined up and symmetrically placed at 120° to each other (Fig. 9.6(b)). The open centre should be roughly the same size at photographic scale as the floating dot so that the three arms are seen extending beyond it; clearly this makes it very easy to place the dot precisely in the centre. The ground size must depend on the scale of the proposed photography and on the instruments to be used, but, as an example, for 1/10 000 photography using a second- or first-order instrument a dot 50 μm in diameter (0.05 mm at photographic scale) will cover an object 50 cm in diameter on the ground so the arms should be 2.5–3 m long and about 30 cm wide if they are to be white or yellow. Light coloured objects expand on the emulsion by halation (see p. 230) so they can be relatively narrow if light, but they must be a good deal wider if they are dark on a light background. In general light coloured marks are best. Paint, lime, sheets of cloth, or wood or thin corrugated iron painted white form useful marks provided that precautions are taken to ensure their safety and that holes are cut or punched in usable materials. Less elaborate marks such as round splashes of paint, lime, or acetylene sludge can be used for the points which are to be fixed by the photography rather than being used to control it.

What is clear from all this is that whenever air photographs are to be used in any operation where features, whether identifiable on them or not, are to be fixed on the ground and a map produced, early cooperation between the scientific user and those who are mapping from the photographs can pay very handsomely in the savings in time and expense. The following points should be particularly borne in mind.

(i) If the area is small and overseas the considerable expense of flying out a land surveyor for perhaps only a week's work can be saved if the field scientist can learn to put in and identify satisfactory ground control, but it is advisable to carry out a trial run in his home country first and he must be shown what is required by the photogrammetrists doing the air survey.

(ii) With good photography, the right instruments, and good ground control, the photogrammetrist can produce an accurate map at scales up to eight or ten times that of the photography, so that each overlap will produce a square metre or so of map; he can contour with an accuracy of half the contour interval which can be equal to about 1 part in 2000 of the flying height.

(iii) Photogrammetry and field survey are two halves of the same technology, and while the former is normally used for detail mapping and the latter for controlling and checking it, their roles can be reversed and photogrammetry can produce a framework of control for detailed survey on the ground of features, including sampling sites, that do not appear on the photographs.

(iv) Close liaison should be kept from the earliest stage in planning such a survey between the field scientist and the survey or photogrammetric organizations taking and handling the air photographs.

Locating airborne geophysical surveys

Preliminary geophysical reconnaissance over large areas is often done from the air because of the greater speed of operation, with ground checks being made on anomalies which appear to be promising. Various instruments are carried, either on board or trailing below the aircraft at the end of a cable. Flying is done at different heights for different purposes, but sometimes at only 1000 ft (300 m) or so above the ground, the so-called *contour flying*. Clearly, if the results are to be correlated with interpretation from air photographs supplemented by ground observations and specimens, the position of the aircraft must be accurately known at all times. It is doubtful whether this can be achieved with sufficient accuracy or continuity with conventional navigation, although modern automatic inertial methods may be able to provide the necessary precision. An easier and so far successful technique has been to use a small vertical *spotting* camera taking photographs at regular intervals with the times being recorded with the geophysical observations. In most geophysical contracts of this kind normal small- or medium-scale air photographs will also be included, unless they already exist, and the positions from which these spotting photographs were taken can easily be determined by comparison with this main photography.

The cameras used are specially designed with very large magazines to take a thousand or more exposures of 35 mm film. The *Vinten* camera is one type. They have a wide-angle lens, with a focal length usually of about 25 mm (1 in) and thus produce negatives at scales one-sixth of that of ordinary 6 in photography taken from the same altitude. This is essential so that a sufficiently large area can be covered by each photograph to provide a basis for comparison with the main photography, which is likely to be at about 1/40 000 scale. From 1000 ft (300 m) the special camera negatives will be at 1/12 000 scale, and reduced or direct contact prints will be near enough in scale to locate them on the main photography. They are then marked on this and incorporated in whatever kind of map is made from it — a controlled mosaic, a proper photo-map, or a line map. The track of the aircraft is then interpolated between these points which are usually a few centimetres apart on the plot.

These cameras can also be used to check on the altitude of the aircraft above the ground, which may have an important influence on the geophysical observations, especially in contour flying. It is likely that a *radar altimeter* will be used and there may be a continuous record of the altitude in the geophysical records, but a check on this is always useful. This can be achieved directly by comparing the scale of the small photographs with the scale of the photo-map, provided that the focal length of the small camera is accurately known. The manufacturers should be able to supply this, but if there is any doubt the scientist can check it by calibrating the camera in exactly the same way as for a camera used for terrestrial photographs as described in Chapter 3. If the geophysicists are operating before the main photography has been taken, they can carry out their own check on the altimeter by photographing a pair of white sheets set out on the ground at the airfield with the aircraft flying straight and level over their midpoint at a predetermined height recorded by the altimeter. At an altitude of 1000 ft (300 m) the width covered by a 35 mm film taken through a 25 mm lens should be of the order of 1500 ft (450 m), and so the sheets should be about 1000 ft (300 m) apart.

A possible alternative is to photograph clearly marked existing features such as the sides of the runway and measure the width of this.

PART IV
The use and production of maps

10. USING EXISTING MAPS

General characteristics

The two most important differences between maps and rectified aerial or satellite photographs are first that the map is always out of date by the time it is published while the photograph is up to date on the day it is taken, though not necessarily of course by the time it reaches the user, and secondly that on photographs all surface features large enough to be visible will appear while the map only shows those features thought worthy of interest by the cartographer.

It should be remembered that this is even true of photo-maps printed in colour lithographically from the original photographs, since decisions on interpretation have to be made by the cartographer in converting the various shades of grey in black and white photographs to a series of colour plates with, for example, yellow for sand, light and dark green for scrub and forest, blue for water, brown for rocks and so on. Although this makes a very attractive and readable map for the general user, it may be less useful for the field scientist who has then had a possibly misleading barrier placed between himself and the original imagery of the photographs. All maps fall into one of two categories:* (i) *first-generation* maps which are compiled directly from survey material (field observations, aerial or satellite photographs, or a combination of these); (ii) *second-generation* maps which are derived from other usually larger-scale maps. Good maps include in the margin a small diagram to show how they were compiled including details and dates of different sources. All maps depend ultimately on an accurate framework and on less accurate detail survey based on this. Since 1945 most first-generation maps at scales of 1/10 000 or smaller have been compiled mainly from aerial photographs, with some field work to provide the framework and to fill in names, road classifications, and administrative boundaries. By the time it has been fair drawn from the original manuscript compilation, colour separated, and printed, a map is usually at least a year out of

*Maps can also be divided into *topographic*, showing the shape of the ground and the features on it; *cadastral*, showing legal property boundaries; and *thematic*, showing the distribution of other information such as population densities.

date. In projects where maps are being specially made for scientists photocopies of the photogrammetric plots may be available before fair drawing and should be used as soon as they are obtainable and possibly before they have been ground checked and revised. Maps of developed and underdeveloped countries are in general different in scope and origin and therefore in their characteristics. In developed countries, such as Britain and northern Europe, medium- and large-scale mapping began several centuries ago and gradually improved in scale and accuracy, particularly in the more built-up areas. The problem now is how to keep these maps up to date; while they were accurate at the time of compilation they may now fail to show even major developments such as new motorways or housing estates since the latest editions of some sheets may be 20 years old. The dates of the original map and of any revision are normally shown in the margin.

Overseas the picture tends to be quite different, especially in the less-developed countries. In these there are virtually no large-scale maps except of the major towns or a few development projects, or specialized title (*cadastral*) plans of property boundaries which are often not on any grid or part of a proper map series and show only the boundary corners and the bearings and distances between them. The rest of the country is usually covered by combinations of the following four kinds of map:

(i) Maps at scales of several inches to the mile (1/250 000 or smaller) compiled from field surveys carried out in the early part of the century by European officers who observed the main framework and supervised teams of local surveyors surveying the detail by plane table or compass traverses. Some of these show contours, but none approach the accuracy of detail and contouring of a modern photogrammetric map.

(ii) Maps at scales between 1/100 000 and 1/1 000 000 that are usually based on widely spaced astronomical fixes and compiled by slotted template assemblies from small-scale vertical, or sometimes tri-camera, photography taken before, during, and immediately after the Second World War. These are usually of the desert and semi-desert countries of north Africa and parts of the Middle East. They have no contours but are planimetrically accurate and detailed, and sometimes indicate relief by hill shading, by form lines, or by reproducing the texture of the photographs.

(iii) Accurately contoured maps at 1/100 000, 1/50 000, or 1/20 000 compiled in three-dimensional plotters from vertical aerial photography and based on an accurate framework. Most of these were made during the last 30 years by local survey departments, or more commonly by large agencies such as DOS* (UK), IGN (France), or USGS (USA), or on contract by one of the larger air survey companies such as Huntings, Clyde (UK), KLM (The Netherlands) or a number of American and Canadian firms.

(iv) Maps compiled directly from satellite imagery.

Although maps of the first type are now out of date and inaccurate, and have usually been replaced, they do show the cultural detail, i.e. the names and positions of villages or tribal areas, administrative boundaries, and the names of rivers and mountains, that was accurate at the time of compilation because the surveyors were working slowly over the ground and had the time and the local knowledge to collect this sort of information. They can therefore be useful to supplement later photogrammetric maps which were often compiled with only the most superficial ground check or addition of names; in many cases these were simply copied from the original ground survey maps making the best fit by guesswork. Thus with all their faults the old ground survey maps often provide the best picture of the social structure and settlement patterns of the country at the time they were made, and the names shown on modern maps must be treated with reserve until one can check how thoroughly field completion was done, either by the mapping organization itself or by the local survey department and administration as their contribution to a joint project. Field completion is a tedious and expensive process whose cost and time are difficult to predict; it delays publication so that it is not always done.

Because of difficulties over security, or simply of a failure to keep proper records, it is increasingly common to find that even when quite good maps have been made from air photographs they are not available, and so maps are sometimes compiled from satellite photographs of the same areas. They are inevitably of lower quality than maps made from air photographs at a much larger scale, but their depiction of natural features is more accurate and detailed than on the earlier maps made on the ground. The temptation to use satellite photographs

*Directorate of Overseas Surveys; Institut Geographique National; United States Geological Survey.

is strong because they are cheap to obtain and to compile, but it should not be allowed to divert any field scientist from making strenuous efforts to obtain copies of the best available aerial surveys, not only by consulting the local government concerned but also by enquiries to the various aid agencies and commercial companies which might have been involved. Clearly it is easier to obtain documents once you know that they exist and exactly what form they take. The main disadvantages of satellite photographs and maps made from them are that, because of the great altitude from which they are taken, they cannot show contours or details of many manmade features. It seems unlikely that better resolution than 30 m will ever be available generally — whatever NASA and defence organizations may be able to achieve — because it would be politically unacceptable; a sharper resolution than this would show military as well as civilian structures.

It is clear from the above that no map is up to date and many are inaccurate in one respect or another. In the remainder of this chapter we first describe grids and graticules, and then discuss different types of modern map in more detail. In each case we describe what to expect in the map and how to use it both for locating oneself on the ground and as a basis for mapping features of scientific interest which do not appear on them. The five map types are as follows:

 (i) large-scale plans in the United Kingdom;
 (ii) large-scale maps overseas;
 (iii) medium- and small-scale maps compiled from air photographs;
 (iv) smaller-scale derived maps;
 (v) satellite photographs and maps made from them.

Grids and graticules

The earth is very nearly spherical in shape since its north–south diameter is only 1 part in 300 shorter than its diameter at the equator, and in this section it will be treated as a sphere. Only small portions of a sphere can be accurately represented on a flat sheet of paper; in the case of the earth the limit is an area of about 30 km × 30 km. To represent larger areas, as on atlas maps, it is necessary to use one of a variety of *projections* in which the curved surface is shown as flat in such a way that some distortion is inevitable; for example, areas may be accurate but then shapes are not, or shapes are locally correct (as in the Mercator projection used for charts) but areas are not, which is why on world Mercator charts Greenland looks as large as South America. Since this book is for field scientists making their own maps or adding field details to maps made by others, it will not describe projections or their principles and mathematical background. Over relatively small areas and at large scales all the projections used for these give virtually the same result, and the differences between them are less than the distortions caused in a paper map sheet by changes in temperature and humidity. However, for purposes of reference the scientist must understand the principles behind and the differences between the graticule and the grid. Both are systems of lines by means of which a unique set of co-ordinates can be given for any point on the earth's surface and thus define its position.

The *graticule* is a lattice formed by two sets of lines. The *meridians* of longitude run north and south and pass through the two poles, while the *parallels* of latitude run east and west and are parallel to each other and to the equator. On the earth's surface, and therefore on large-scale maps of small areas where shapes are correct, these intersect at right angles. This is a universal system, and virtually all countries have now adopted the version in which the parallels are numbered in sexagesimal (360° to a full circle) degrees from 0° at the equator to 90° at the poles, while the meridians are numbered east and west from the Greenwich (London, England) *prime meridian* 0° until they meet in the Pacific at 180° both east and west. A few maps may occasionally be seen where *grades* (400^g to the full circle or 100^g at the poles) are used or where a prime meridian other than Greenwich has been used, but these are increasingly rare. The two sets of lines, those running north and south and those running east and west, differ fundamentally, mainly because of the rotation of the earth and the importance to navigators of the altitudes of heavenly bodies at different times of the day. Meridians are *great circles* with their centres all at the centre of the earth; thus on any large- or medium-scale map they will appear as straight lines. Parallels are, except for the equator, *small circles*; near the poles they become very small and in fact at the poles become points, so that in high latitudes their curvature is appreciable even on large-scale maps; they are not straight lines. All points on a parallel of latitude will see the sun at approximately the same altitude above the horizon at the same local midday, but this will be at a different moment of time for each point along the parallel. In contrast all points on a meridian (except those too far north or south) will see the sun due south or north and at its highest altitude (and lowest also at midnight in high latitudes in summer) for the day at the same moment, but the altitude will be different in different latitudes and so will the direction, whether north, south, or overhead.

Clearly the graticule is world-wide and universal; every point on the earth's surface has a unique reference in latitude and longitude in degrees and minutes (to roughly the nearest mile) or if required in seconds also for greater precision. All good atlases include a list of the place names shown on their maps – a *gazetteer* – and this lists not only the maps on which they appear but also their latitudes and longitudes. However, the graticule has disadvantages, mainly because of its irregular shape, which makes it more difficult to calculate or scale off precise co-ordinates correctly, and also because these are not in a decimal system. Thus for most purposes, particularly military ones but also for scientific use, a *grid* is used. As already explained in Chapter 6 the scientist using instruments accurate enough to require calculation and plotting of co-ordinates must plot these on a grid, which in many cases he simply draws up as a series of squares on his map sheets, starting from an arbitrary datum and aligning them approximately north and south by compass or by reference to an existing map (or by astronomical observations for azimuth). Alternatively, if the survey is based on an existing framework, he can use the grid on which the co-ordinates of its points have been given to him. Grids differ from the graticule in several ways. First of all, with one exception, they are not world-wide but are local and designed for a special purpose. This design may be, as in the case of the United Kingdom's national grid, to suit a country isolated from its neighbours by sea, or it may be to suit a special project and sometimes grids are oriented askew (i.e. not north and south) for this purpose. The one world-wide grid is the Universal Transverse Mercator (UTM) which covers the whole earth between 80° latitude north and south in a series of *zones* each of which is 6° of longitude in width. Within these, as within the United Kingdom national grid system, the restriction in longitude means that the inevitable distortions due to representing a curved surface on a flat plane are minimized so that the maximum *scale error* anywhere (as between a distance calculated from grid co-ordinates and measured horizontally at sea level) is less than 1 part in 2000 which is less than paper distortion. This sort of difference is unlikely to bother a field scientist unless he has gone beyond the scope of this book, in which case he should use a more advanced textbook for calculating his co-ordinates.

The obvious advantage of the grid is that the lines defining it are straight and at right angles on the map; this makes calculation of co-ordinates from bearings and distances very simple, and facilitates accurate references by using grid squares. However, it must be appreciated that, since the meridians are not parallel to each other except at the equator, outside the lowest latitudes it is only at the centre of a grid zone that the grid lines run exactly north and south, and the further one is from the equator and from this centre line the greater is the *grid convergence*, i.e. the angle between true north and grid north. This is shown on the margins of all gridded maps, and in the United Kingdom for example it may be as much as $5°$; clearly in the west of a grid zone grid north will be west of true north, and in the eastern half east of it (in the northern hemisphere). In general, the smallest-scale maps do not bear a grid at all, medium-scale maps tend to have sheet lines based on the graticule but bear a grid across their face, while large-scale maps and plans usually have sheet lines based on the grid and also carry the grid on their face. However, in nearly all cases the graticule values of the corners, and *graticule ticks* for values of latitude and longitude such as minutes, are shown in the margins, and sometimes crosses appear on the face of the map so that the user can, if he wishes, find not only the grid reference of a point but also its latitude and longitude. Charts used not to bear grids at all, but with the closer connection between sea and land arising out of the development of minerals under the sea, this is beginning to change.

All gridded maps normally give instructions in a box in the margin with an example of how to give a grid reference, and these should be followed. All grid co-ordinates are given from a datum or origin to the south-west of the grid area, so that co-ordinates will always be positive when measured eastwards and northwards. Therefore the procedure is to write down first the easting and northing of the *south-western corner* of the grid square in which the point appears, and then add to these the intermediate distances inside the square from its western and southern sides. This is done to the nearest 100 m in medium-scale maps, and to the nearest 10 m or even single metres in large-scale maps. Sometimes (as in the United Kingdom) the whole grid area is divided up into 100 km squares each with a distinctive letter or pair of letters, but any special arrangement of this kind is explained in the grid reference box. Graticule references or *geographical co-ordinates* are more difficult because, especially in high latitudes, the 'square' is not a square or even a rectangle, and as already mentioned the parallel north and south sides may not even be straight and in higher latitudes are visibly curved. Thus the latitude parallel and longitude meridian of the point have to be drawn through it by eye until they cut the sides of the graticule 'square'; their distances from the southern and western ends of these have then to be measured and converted to minutes (and if necessary seconds) either by calculation or by comparison with the divisions of the graticule which are shown at the edges of the map. In high latitudes the side of the 'square' nearer to the pole will be appreciably shorter than that further from it owing to the rapid convergence of the meridians as they near the pole.

Large-scale plans in the United Kingdom

The basic mapping scales in the United Kingdom are (i) 1/1250 for towns (approximately 50 in to the mile), (ii) 1/2500 for rural areas (approximately 25 in to the mile), and (iii) 1/10 000 for mountains and moorland (approximately 10 in to the mile).

The Ordnance Survey's annual report shows the areas for which each series is the largest scale available and is the basic map. Areas covered by the larger scales are also covered by derived map sheets at smaller scales and the 1/10 000 series covers the whole country including towns. The work has been carried out by the Ordnance Survey over a long period of time, so that in some areas, particularly rural ones, the basic maps are still based on field surveys carried out 100 years ago. However, all the 1/1250 series, and most of the 1/10 000 maps of Scotland, have been completely resurveyed on the ground or from air photographs since 1950 with extensive ground checks for names and other details. The 1/10 000 series is contoured but the other two series only include spot heights at high and low points and changes of slope, so far as the density of buildings makes it possible to show them. Thus they are not so much maps of the ground itself as plans of the manmade features on it, including not only buildings and roads but also hedges and forests. Their principal use, and the justification for the expense of keeping them up to date, is to record property in cartographic form as an adjunct to written descriptions, as a basis for recording ownership and other rights in private conveyances and land registration,* for implementing the planning regulations which now control almost all building in this country, and for local authorities responsible for services such as police, drains, water and power supplies, fire fighting etc.

Litho-printed fair drawn sheets still form the majority of the material on sale, but sheets of the 1/1250 and 1/2500 are now being issued in other forms. Copies can be obtained on transparent film from which the user can make contact copies by the dyeline process, although copyright fees have to be paid. Sheets are also issued in digital form on tapes or discs for those equipped with the expensive automatic plotters to reproduce them. At present this facility is mainly used by large organizations like county authorities or state corporations, but developments in digital computers are so rapid that in a few years most local authorities and many university geography or other departments may be able to reproduce required parts of the large-scale plans quickly at any required scale and to any special specification in which, for example, only certain features are shown. The description and use of digital data are beyond the scope of this book.

At present the most up to date and generally useful forms in which this map material is issued are known as SUSI and SIM. Both of these are based on the master tracings in the Survey's field offices which are kept up to date by *continuous revision*. This is not completely comprehensive and does not include every minor change in rural areas. SUSI is a contact copy of the master trace prepared by the Ordnance Survey staff and is obtainable from their offices; SIM is produced by enlargement from a microfilm by agents who may vary considerably in their skill in making a clear and undistorted copy. It is cheaper than but inferior to SUSI. The master trace shows the original parts of the last fair-drawn edition of the sheet, with later changes added by deleting buildings or roads which have been demolished and adding what is new; these are recorded in neat surveyor's hand and not fair drawn by a professional draughtsman. Simple hand-drawn symbols such as dots are used to indicate buildings rather than the machine-made *stipple* fillings (fine regular patterns of dots) used on fair-drawn plans; the result is neat and comprehensible but not a work of art

*Public boundaries (e.g. parish boundaries) but not private ones, are *mered* or defined on the ground by reference to a physical boundary feature by Ordnance Surveyors as a statutory duty, assisted by representatives from both sides, and full records are kept at the Ordnance Survey in Southampton. See Booth, J. R. S.: *Public boundaries and Ordnance Survey 1840-1980*. Ordnance Survey, Southampton (1980).

like a finished fair-drawn map. The SUSI and SIM material is not more than a year out of date, apart from minor and widely separated changes which are very uneconomic to survey. Although they have a good intelligence system to advise them of changes on the ground, the Ordnance Survey officers are always glad to know of any changes or errors on their maps.

There are still a large number of sheets in the 1/2500 series based on the old 'county' maps (which were on local grids) recompiled onto the national grid by methods which at times result in errors as large as 5 m (2 mm at map scale). This is an 'absolute' error (see p. 80), and 'relative' errors of this size are unlikely to trouble a field or air surveyor unless he is working accurately over long distances. Where the map has been accurately surveyed, as has happened to all the 1/1250 sheets, errors in the features shown at publication date should not exceed what is normally plottable, i.e. 0.5 mm on the map, corresponding to 0.6 m on the town plans (2 ft), 1.2 m (4 ft) on the 1/2500 maps, and 5 m (17 ft) on the 1/10 000 series, but of course the map may still show features which have been demolished and fail to show others built since it was surveyed. Whether one uses the published (but probably more out of date) printed sheet or SUSI or SIM should depend on how much change has occurred in the area of interest since the sheet was last published, whether the sheet is to be used extensively in the field or only in the office, and the funds available. It should be noted that to buy one copy and then make further copies from it may be infringing the copyright rules. The problem should be discussed with the nearest Ordnance Survey office.

Two stages are required for positioning oneself or for surveying a boundary or feature not shown on these or on any large scale plans:

(i) the selection of a framework of correctly and precisely identified features already shown on the map;
(ii) measurement from these by simple means such as compass bearings or taping, or even careful pacing, to the points or positions which are to be fixed.

Of these two stages the first may well be the most difficult, particularly on an old or out-of-date map, while stage (ii) is similar to the detail survey based on a normal framework. The first stage requires commonsense and a deep distrust of all data until it has been checked. The best way to check that features like building corners or hedge or wall junctions have been correctly identified and have not changed is to measure distances or bearings between them. Since the local errors on these large-scale plans are 1 m or less, the agreement between measured or derived short distances and those scaled off the plan should be so close as to preclude a coincidence, except in repetitive landscapes like housing estates or factories where one may be identifying the wrong house or even the wrong road.

The features providing the framework must surround the area as well as provide points within it; it is bad survey practice to work from one side only. They should be close enough to make the survey work required to fix oneself from them quick and sufficiently accurate; a wider spacing is possible using compass bearings or a tape than with an optical square or pacing which are only accurate over short distances. They should be reasonably accessible without crossing too many ditches, thorn hedges, or walls, or having to ask permission from too many house or farm owners; features along a road or public footpath have obvious advantages. A feature should be sharp enough to be identifiable on the plan to within about 0.5 mm. Suitable features are building corners, wall, or fence (but not movable electric fence), junctions rather than hedges, and bridges or culverts over a stream – in fact almost exclusively manmade rather than natural features. On maps at the two larger scales buildings are shown almost exactly in their correct ground plan shapes, and so one can see easily if one has been extended, such as a new wing to a house; but there has to be considerable generalization on maps at 1/10 000 scale and many minor corners or juts in buildings are not shown. In such cases the best corner to take is the one with the longest length of straight wall on each side. A stream in a steep valley obscured by trees is unlikely to have changed since it was surveyed, but the field or photogrammetric surveyor may not have been able to survey it exactly if the trees and thickets round it were too dense. Thick hedges accompanied by banks and ditches tend to be shown on Ordnance Survey maps by a single line.

Although the detail survey from these points will be much the same as from one's own framework in an unmapped area, it may be possible to choose so many points in a developed area that the scientist can use the system of straight lines much favoured by Ordnance Survey revisers (e.g. when adding new houses built in a field). Points are chosen so that straight lines between them pass near the object to be surveyed, and its position is then easily determined by measuring the perpendicular distance from it to each line and the distances along this from one end. For accuracy and even longer distances one should use an *optical square* in which two objects either at 180° or at right angles can be seen simultaneously by a system of prisms; this ensures that the detail is plotted on the line joining them or on a line perpendicular to it.

The 1/10 000 scale maps may be compilations from the larger scales or from direct surveys, and the whole of the 1/25 000 series is compiled from larger scales with only a few features surveyed directly to bring them up to date. The larger-scale maps are likely to be more accurate and less generalized, but if a scientist's results are going to be published at one of the smaller scales, it may well be a waste of time to survey them onto the larger-scale plans and then have to reduce them. The extra cost and handling time of using large-scale plans could be very great since the number of sheets of equal size increases as the square of the ratio between the two scales. All these points should be considered in deciding at what scale to do this basic mapping.

Large-scale plans overseas

There is seldom a comprehensive large-scale topographic survey in the developing countries anything like that of the United Kingdom. Large-scale plans or maps that exist on scales at 1/10 000 or larger are normally (i) town plans, (ii) cadastral plans of property boundaries, or (iii) special project maps. These are of very different value to the scientist wanting to locate or set out features not shown on them.

The large-scale *town plans* are much the same as those in this country, although when they have been compiled from aerial photographs some ground detail may be incomplete. They will be based on a proper framework, and where the detail has changed a great deal the scientist should use the points of this – if they can be found. Many town plans were made for planning future extensions and so include considerable areas outside the built-up area. In these areas there may be relatively little 'hard' detail such as permanent buildings or walls, and the framework points will be even more valuable for controlling and locating new survey work; they may also be easier to find. Where orthophotographs have been made bushes or small trees can be used. With line maps the scientist may find it worthwhile to obtain the original air photographs and make a radial line assembly from contact prints or enlargements.

In judging whether the *cadastral plans* can be of value, probably the most important factor is whether they are individual and unconnected or whether the local survey department has managed to fit them together onto the grid. In many overseas countries such as those in North America, Africa, and Australasia, grants of land to settlers were marked by corner pegs or monuments and surveyed, mainly by private surveyors, onto individual title plans or set out in grids as in the USA or Canada before a comprehensive national framework or map series was available. Not only are many of these cadastral plans or blocks of them often not connected together, but in most cases they show only the boundaries and monuments and the bearings and distances between them, or a list of their co-ordinates; buildings and other more obvious landmarks are often not shown at all and certainly not kept up to date. A national survey framework, medium-scale topographic mapping of the country, and large-scale topographic mapping of the towns came much later. In the Middle East the Ottoman administration established written records of the land parcels in settled areas, mainly in the towns and river valleys, and these were marked by corner stones or walls. Survey departments then mapped these comprehensively at large scales.

Except where the cadastral plans have been produced on a regular basis by the local survey department, it is unlikely that they will be much use to the scientist unless the area to be investigated comprises a relatively small number of individual title plans, e.g. of one or two large estates or farms, or is on a grid system as in North America. Scientists tend to be brought up on topographic maps; in Britain they may never have seen a cadastral plan, so that its complex of distances and bearings and absence of all manmade features other than survey monuments comes as a shock. In the field, for example along the Nile flood plain or on terraced hillsides, the cultivation or terrace pattern may bear little relation to that of ownership; there is probably a complex system of leases and co-operative management, with the choice of crops following soil rather than ownership boundaries. In such areas features on air photographs or what one sees on the ground may differ completely from the property boundaries and hence from the accurate cadastral plans, and the best solution is to consult local farmers who usually know who owns which plot, what its parcel number is, and where the boundary marks or 'pegs' are. Having found some of these the scientist should check them by bearings and distances between them and look for others by extending this system to obtain their approximate positions. These corner 'pegs' and the permanent marks of any reference framework vary from stones and mud mounds to concrete or metal pegs, or in central and southern Africa to pillars 3 or 4 ft high (sometimes even higher) with permanent beacons on top. Unfortunately, the more up to date the cadastral system, the more likely it is that any scientific surveys required will also have been completed.

Project surveys vary a great deal. A 1/20 000 scale contoured map may suffice for a large reservoir, while for the dam site itself or for a bridge scales as large as 1/500 have been specified for calculating volumes of earthwork and concrete. Contour intervals also vary with the steepness of the ground, and they may be as small as 0.5 m where the carriage of water or sewage is involved; they tend to be closer than in national map series. Project surveys are usually commissioned by civil or structural engineers, who may also let contracts for soil and other capability surveys. Ground or air survey firms are often employed as contractors and some of these will establish a properly marked framework even if not required to do so in the contract. If temporary wooden pegs were used these will have disappeared, and paper maps are a poor substitute for precise and permanent co-ordinated marks in the ground. If the scientist is involved at an early stage and will have to survey his own results onto the map, he should press for the framework to be permanently marked; the extra cost of doing this is very small, but few survey firms will pay for it out of their own budget if they cannot persuade the client to do so. This is particularly important if the survey covers what is at present an uninhabited or undeveloped area with very few permanent manmade features such as brick or stone buildings. Again, if orthophotographs are made much of this difficulty disappears because of the availability of suitable small natural features appearing on them.

Project maps often use a local datum and grid because of the cost of connecting the survey to the national framework. The scientist involved with consultants should press for this connection to be made especially in surveys for development where the maps and the framework should be designed for a very long life and be available later for many purposes including records of ownership as well as for the initial engineering design and setting out. Local datums and grids, and grids and sheet lines oriented askew (i.e. not north and south) are particularly common in long narrow surveys for roads, railways, or pipelines where the engineer wants to handle as few map sheets as possible and relies on a single traverse running through the project. Since the survey itself is so narrow, its main value to scientists may well lie almost entirely in the points of this traverse. Areas of interest alongside the new line of communication can be surveyed using these traverse points as the starting control. Air photographs used for such surveys will have been flown in single or double strips along the proposed trace and not oriented north–south or east–west; it may well be worthwhile to obtain copies of these and use them for detail surveys with radial line methods.

Medium- and small-scale surveyed maps

Virtually all first-generation medium-scale maps (on scales from 1/25 000 to 1/200 000) are now made accurately from air photographs. While specifications vary a great deal, the general features of such maps are very much the same. In medium-scale mapping it is uneconomical to use photographs on much larger scales than about 1/40 000; with superwide-angle cameras or jet aircraft, or both, the scale may be as small as 1/100 000. On such photographs manmade features like small villages and tracks or isolated buildings may scarcely be visible at all. To the map user these may be important features, but unless the map has been properly checked on the ground they and their names and classifications, for example whether the tracks can be driven over or not, may well be wrong or missing altogether because in inaccessible areas the work of ground checking may not have been done at all. Thus while the depiction of natural features and of the contours (if shown and if the tree cover is not too dense) should be accurate, a number of manmade features may be omitted. Where maps are still in the course of production, copies of the photogrammetric plots which have not been ground checked or fair drawn may be almost as useful as finished maps, and the field scientist may find useful a combination of the latest photogrammetric plots and copies of the older (and probably smaller-scale) maps based on early ground surveys. Even if changes have occurred the local people may still remember the sites and names of old villages shown on them.

A very useful feature of some maps prepared from air photographs (particularly by the DOS), and of most photogrammetric plots, is the addition of the principal points of the vertical air photographs. These make it easy to add detail (such as vegetation or soil boundaries or even landmarks like trees) not shown on the maps with precision by using radial line methods. This use of the air photographs may be the only way to locate oneself or other features accurately in less developed areas without using quite elaborate ground survey techniques for breaking down the existing survey framework, whose points may be far apart and inaccessible. Very large areas of tundra or semi-desert in North America, Africa, and Australia have been mapped at scales of 1/100 000 or less by slotted-template assemblies of vertical (or vertical plus oblique) photographs without adding contours. Stream beds, lakes, glaciers, dune areas, and other features can be accurately shown, but hills or mountains are only indicated by such devices as tracing the ridge lines, approximate form lines, or hill shading (as described on p. 273). Hill shading can be done by the cartographers from a study of the photographs, or in some cases the shading on these can be reproduced directly on a brown or grey plate as a background to the detail and names. These make it possible to locate oneself by map reading and indicate the relative steepness of slopes and heights of neighbouring mountains, but it will not be possible to make any accurate estimates of heights or slopes from them. Maps of heavily forested areas made from air photographs may give a misleading picture of the topography because only the tree canopy can be seen and not the ground, and even quite large streams may be invisible. Moreover, because trees near streams obtain more water and therefore grow larger, their canopies may actually be higher than those of trees further from the water, and thus the bottom of the valley appears to be above and not below its sides.

Small-scale derived maps

1/250 000 is the smallest scale at which maps are normally prepared directly either from ground or aerial surveys, and most maps being produced at smaller scales are what is called *derived* or *second-generation* maps, that is they are compiled from original surveys at a larger scale. Derived maps, especially as the scale becomes smaller, have two main characteristics: first they tend to be even more out of date than maps based on original surveys because they have to wait for these to be completed for the whole area covered by one or more sheets of their own series, and secondly they vary in reliability even over a single map sheet depending on the accuracy of the first-generation maps from which they were compiled. However, there are still considerable areas of the world where small-scale maps at 1/250 000 or less are still the only ones available to the public or even to visiting scientists invited to do work in the country concerned. Most of them include in the margin a compilation diagram and list of the data used, but further enquiry may be needed to establish the reliability of this. The style of drawing should give some indication of the accuracy of different parts of a map; stream lines, contours, and roads and tracks are usually shown with much smoother curves, or even pecked, in areas where they have not been properly surveyed, while in surveyed areas they will be shown firm and in full detail. There are exceptions since symbols used for dunes and mountains are sometimes drawn freehand from aerial reconnaissance and thus look accurate and detailed, but ground checks will show them not to be so.

On the older small-scale maps compiled from travellers' journeys, the features along each journey may be shown in detail and accurately, but those away from it may have been mapped only from a few bearings or even from information supplied by guides. Many of the great exploratory journeys were accomplished by travellers interested in the inhabitants rather than the topography, and they were, moreover, often obliged to take the minimum of equipment and carry out their survey observations clandestinely. Astronomical observations for position were often impossible, and distance measurement often depended on estimates of walking or riding speeds. Thus errors of several miles could occur in the long stretches between points whose positions were accurately known. During World War II some 1/1 000 000 scale series were compiled from maps already prepared separately by different organizations using much of the same travellers' records; at the common sheet edges comparatively large towns disappeared completely and in one country a strip some 10 miles wide appeared on both the adjoining sheets. These had to be reconciled in the new series by guesswork without any further information on how to position correctly the missing or duplicated detail. Therefore, in using this type of map in the field, the scientist would be wise to try and locate himself from features whose positions are most likely to be correct; examples are the coast line, particularly if based on British Admiralty Charts, major towns, and major roads or railways or other engineering projects.* Away from these he may find that his own dead reckoning navigation, based on careful compass or sun compass and car milometer traverses, gives much better and more consistent results than the map. In areas with mountains and good visibility

---

*Older roads or railways may however not be accurately mapped. The railway from Egypt into the Sudan, built before 1900, was found in 1955 to be four miles out of position at one point. This had no effect on the running of the trains.

it may be possible to use a combination of dead reckoning and a plane table or theodolite survey from which mountain peaks can be fixed or, if their positions are found to be reliable, from which the scientist can fix himself. Although it is normally used at medium scales, a plane table can be used successfully at scales as small as 1/500 000, particularly if visibility is good and travel is by car or truck between hills from which a good view can be obtained after a short climb. Up to a thousand square miles of desert a day was mapped in this way during the Second World War. Whether a plane table is used or not, it pays to plot one's traverses and any detail fixed from them in the field so that any inconsistencies can be sorted out on the spot.

In working on maps of varying reliability in this sort of way the normal survey principle of checking everything and taking no data for granted until proved is even more important than usual, and it is as well to remember that the reliability of such maps is often inversely proportional to their artistic merit and cartographic sophistication. All the scientist's route traverses must be closed at both ends onto reliable features, and it is worth investigating what is available in the form of astronomical or satellite fixes, even if they are not shown on the map. In desert areas the sites of these, being camp sites, can often still be recognized. Resections from distant mountains should use more than three well-spaced rays, and internal checks like closing one's own traverses on themselves or each other should be applied as frequently as possible. Frequent checks should be made on magnetic variation using bearings along tracks of known accuracy, from the sun using azimuth tables, or from the Pole Star in suitable northern latitudes. Unless a sun compass is used or the magnetic compass has been properly 'swung' in the vehicle (by pointing this on a series of known bearings and noting the difference from the compass reading), all bearings should be taken well away from it and, near railway lines, well away from the tracks. Where a winding track or course has to be followed and two or more vehicles are available, the navigator should travel in the rear one and take bearings to the front of the convoy at least half a mile ahead; this will smooth out the minor turns in the route which are too small to be plotted and difficult to average. Plotting should be done carefully because, even with this precaution, each leg of the traverse will only be a few millimetres long at small scales and bearing errors will easily accumulate.

At small scales it seems unlikely that features (other than topographic ones) to be mapped by the scientist can be identified except by visiting them on the ground, and so the survey work is more a matter of simple navigation, with the traverses planned to follow soil or other boundaries rather than as a framework from which they can be fixed. However, in very open desert country extra bearings to features off the track can also be used up to about 10 km on either side, and in such cases the same kind of techniques as already described in Chapter 2 should be used. Where the existing map proves to be unreliable despite the amount of detail shown on it, the scientist may be better off using a blank sheet of paper, only transferring to this the graticule and useful reference points which have proved to be correct. Clearly the margin between adding detail to maps of this kind and carrying out completely new surveys becomes difficult to define.

Maps from satellite imagery

A completely different form of small-scale map, made from satellite imagery, has become available in the last 10 years. These maps are obtained by direct reproduction, by conversion to a line map, or by using the satellite photograph to revise an existing map. Some of the exaggerated claims that were made initially, for example that satellite imagery could be used for 1/50 000 mapping, have proved to be false, and the general opinion now is that for topographic and locational purposes the largest scale on which satellite photographs can safely be used is 1/250 000 since most of the original photographs were taken on scales of 1/1 000 000 or smaller. This of course applies only to those photographs made generally available; classified photographs are taken on very much larger scales for military purposes. Thus the main use of satellite imagery to the field scientist is in the areas mentioned in the previous section where the best available maps, for whatever reason, are at this or smaller scales and are often unreliable and out of date. In this section we outline how satellite imagery is converted to photographs or maps and how it can best be used, but it will be worthwhile to state first what its main characteristics are.

(i) The scales are small and the resolution of existing imagery is usually not much better than 50–100 m on the ground.
(ii) As a result of this only major forms of human activity can be identified; these are mainly in developed areas where good maps and aerial photography already exist.

258   USING EXISTING MAPS

   (iii) Each photograph covers a very large area, somewhere about 100–200 km square, corresponding to a large number of air photographs even if taken from 30 000 ft.
   (iv) Many satellite photographs are taken at regular intervals of the same area and at the same time of day, so that seasonal changes in tree cover or crops can be seen; all except persistently cloud-obscured areas are covered.
   (v) Geometrically satellite imagery resembles a map rather than an aerial photograph, but while relative positions may be accurate, absolute positions may be wrong if the graticule has been incorrectly positioned.
   (vi) Because of the great altitude from which they were taken, there is virtually no stereoscopic effect and no heights or contours can be measured.
   (vii) Since at present only processing charges have to be met, satellite photographs are cheap.
   (viii) Satellite photographs are worldwide and are no respecters of national frontiers.
   (ix) The imagery was produced by several different bands of radiation including infrared.

How satellites operate and how they produce photographs are described in the books and papers listed in the bibliography. Satellite imagery falls into two distinct classes: the first class includes sporadic but often oblique photographs taken on film during short-term missions such as Skylab, the Space Shuttle, and high-level rockets which are recovered and processed in the normal way; the second class includes all those taken regularly and usually vertically from long-term unmanned satellites where the imagery is transmitted to earth by radio and television techniques. The quality of the first type is likely to be much better for two reasons: the altitudes are lower (100 km compared with 400 km), and the recovery of the film means that the resolution at image scale is comparable with that of an aerial photograph. However, in order to maintain a satellite in orbit for a long time it has to be at an altitude of several hundred kilometers (around 1000 km) to avoid atmospheric drag. Thus on long-term missions the scale of the photographs is smaller, and transmission by radio means that the image has to be broken down into a finite number of squares or *pixels* which are in general a good deal larger at image scale than the grains of a normal air photograph emulsion. A common factor in the orbits or paths of all objects travelling by momentum and without rocket power is that their plumb points on the earth follow great circles or, strictly speaking, spirals because of the earth's rotation; moreover, with long-lasting satellites repeated checks on their orbits make it possible to know their positions and altitudes very accurately, more so than for example with an aircraft on a long passage over unmapped areas. Thus, while contouring is not possible from satellite pictures, special satellites can record height profiles with a radar altimeter.

The most important long-term satellites for earth scientists are likely to be the ERTS or Landsat series, whose orbits have been very carefully designed. In order to cover as much of the earth's surface as possible they are near-polar orbits, i.e. they approximate to passing over the two poles with their plumb points nearly following meridians. They have a period of about 1 hr 40 min, and as the earth rotates beneath them each successive path of the plumb point follows a different path about 20° in longitude further west, so that after 18 days it is back over the same path having covered virtually the whole earth. The orbit is also designed so that each time the satellite covers the same approximately north–south strip it is at the same time of day. The width of each strip is nearly 200 km which gives a small overlap at the equator and of course a progressively larger overlap in higher latitudes; individual photographs along each strip also have a small overlap, but not as much as the 60 per cent found in aerial photographs which can be used stereoscopically. In addition to normal black-and-white photographs, images are also recorded on a number of different near-optical wavelengths such as the infrared. Since specifications change and some recording systems on individual satellites fail, the details are not given here.

In long-term satellites the use of film is impracticable both because of the bulk which would be required and because it cannot be recovered. The pictures have to be transmitted by radio and television techniques which are constantly changing but have one feature in common; they require the breaking down of the picture into discrete elements and the greyness or brilliance of each element or pixel to be described by a number in a scale, so that the transmission can be in the form of numbers or digits. These describe the position of each element by its row and column number (or y and x) and its shade or colour, and convert these to binary numbers so that what is actually transmitted is a series of 'yeses' and 'noes' or pluses and minuses, as in fact is now done with the recording and replay of television series like 'Match of the Day'. Without going into detail it is clear that the ordinary 230 mm square aerial photograph, with a resolution of 5 μm transmitted in this way, would require the details of over 10^9 $(230 \times 200)^2$ pixels to be recorded and sent, requiring a larger number of *binary digits* or *bits* to locate and define each one; thus the problem of recording, transmitting, and processing information in this way are immense. From the user's point of view the three important facts are first that resolution is less good than for photographs reproduced from film, second that relatively cheap film negatives can be obtained which have been processed by the Space Agency themselves, and third that for the best results it is necessary to use tapes or even to receive the radio signals oneself on tape and then reprocess them with more refinement, including such techniques as computer enhancement to increase the contrast. Most scientists, for the present at least, are likely to have to be content with using photographs produced by others.

The main use of satellite data for the field scientist with which this book is concerned, is either as a substitute for, or more commonly in conjunction with, small-scale maps in areas where nothing better can be obtained. If no comparable maps exist the area is likely to be almost uninhabited, and the more this is true and the less the area is covered by cloud the more use will satellite photographs be; obvious examples are the polar regions and the deserts or near-deserts of the northern hemisphere. In general, a scientist wanting to use satellite photography as a map would be well advised to consult either a national cartographic agency such as, in the UK, the Directorates of Military or Overseas Surveys, one of the air survey companies who have had experience of its use, or one of the universities where some departments now have experience of it.

Having obtained the best available cover in this way, probably in the form of a black-and-white enlargement to 1/250 000 or 1/500 000 scale, what sort of features can be recognized on it as an aid to locating oneself on the ground? Unless the photographs have been taken from short-term low-level missions, the imagery obtained at optical wavelenths will be much inferior to that seen on small-scale prints or mosaics made from even the smallest-scale (about 1/80 000) aerial photography in common use, especially since no further magnification is possible and it cannot be viewed stereoscopically. Only major towns, settlement schemes, canals, or roads are likely to be identifiable; individual trees cannot be seen, and though individual fields can sometimes be seen the texture of the vegetation or crop cover will not appear. Forests, swamps, cultivation, or other features can often only be distinguished by different shades of grey, though the relatively wet areas can be distinguished on imagery from other frequencies such as infrared. Rocks and mountain areas will be clearly distinguished from sand or snow, and drainage patterns can sometimes be clearly seen even if they contain no water; coastlines of the sea or lakes will be clearly recognizable, though in flat areas their position may depend on the state of the tide or season. Heavily forested areas are likely to be cloud covered and may therefore never be clear; even if they are the smaller streams are likely to be invisible. In sunshine the relief of mountainous areas, including ridge lines and summits, may well be clearly shown by the shadows and even seem to give a stereoscopic effect when two overlapping prints are viewed in a stereoscope, but this is misleading and no accurate heights or contours can actually be obtained. If line maps are produced it is better to use the actual imagery as a form of hill shading than to drawn formlines.

Satellite photographs form the best location maps in areas such as polar or arid deserts where there is plenty of contrast in tone between mountains and other areas, near coastlines, in an area covered by lakes, or where a well-marked drainage pattern exists. Graticules are usually shown on these prints, but they should be checked against the best available maps at the same scale, comparing features which are easily recognizable on both and which lie in areas shown on the map's compilation diagram to have been well surveyed. Most of the long-term satellite prints have been rectified so that they are true to scale everywhere, and a few check points will serve to locate the whole of one print. However, those taken on short-term missions such as Skylab or from high-altitude rockets may be obliques with considerable distortion; in such cases more check features will have to be located to control the effects of this. Where no recognizable features can be seen, the scientist will have to rely on his own navigation between fixed points as when using inaccurate and unreliable maps; the techniques would be the same as those already described in the preceding section. In flat areas he may have to run a traverse along part of some well-defined feature such as a river or coastline in order to compare its shape with the photograph, since he will be unable to view a sufficient length of it from above. One last point is worth making again: satellite photography has the attraction of cheapness and of the glamour of using a new technology. Its availability should not stop the scientist from making every effort to obtain the best available maps or air photographs of the area concerned, since these may well be much more useful for a proper study of it even though a persistent and irritating investigation may be required to locate them.

11. PREPARING THE BASIC PLOT FOR REPRODUCTION

How maps are produced

Up to now we have been concerned almost entirely with the establishment of a framework of points between which to draw the detailed features to be shown on the map, either on the ground or from air photographs, in both cases by relatively approximate methods. We are now concerned with the preparation of the map itself. First comes the selection, classification, and depiction of the detailed features on the *basic plot*, which may be a field sheet surveyed directly on the ground or a compilation from air photographs. Once this basic plot is complete the scientist has finished the *survey* task; the remainder of the task is *cartographic*. Survey comprises the collection and positioning of the data; cartography† is the presentation of these data in map form, including the production of several copies. Cartography also includes the compilation of maps (usually at a smaller scale) from other maps, but its main function for the field scientist is the conversion of his data to an attractive and clearly understood form (*fair drawing*), and from a single hand-drawn original to a form from which a number of copies can be made by *photomechanical* or *photographic* (dyeline) methods. In this chapter we outline briefly the whole process of map production so that the scientist can understand the factors affecting his drawing of the basic plot, and in Chapter 12 we describe in more detail the process of fair drawing by which he, or a professional cartographer, converts the basic plot into a form which the photoreproduction operator and printer can use. As in surveying, actually to carry out a few simple tasks himself will give the scientist a better understanding of the requirements and attitudes of the professionals than any amount of reading, and it should give him a good idea of the points to look out for when negotiating or monitoring a cartographic and printing contract.

Suppose for the moment that the scientist has surveyed a basic plot including hand-drawn lines, symbols, and names, and showing all the detail and perhaps contours that he wishes to be printed. It is

† The United Nations terminology defines cartography as the whole process of surveying and mapping, but we prefer to follow the normal English definition.

likely that for clarity and to separate individual features he will have drawn the field sheets in more than one colour, and before starting to fair draw them he must first decide how many colours are to be shown on the final map. If he consults a printer he will find that the addition of every colour after the first adds something like 50–100 per cent to the cost of printing a map, and therefore while a multicoloured map is of course both clearer and more attractive than one in a single colour, it is much more expensive. The reason for this is that the *lithographic printing* normally used for maps consists of the transfer by pressure of an image in ink on a plate to successive sheets of paper, each of which is a copy of the map. Only one colour can be transferred at a time, and so for each extra colour beyond the first a new plate has to be made, inked up with that colour, and set up accurately in the printing press. If a large number of copies are to be printed (say over 500), then this may consist of several machines linked together so that two, three, or even four colours can be printed in one operation as the paper passes through the machines; but for shorter runs, which are likely to be more common in the case of maps accompanying scientific reports, a single-colour machine can be used. In this case each extra colour beyond the first requires each sheet of paper to be passed again through the same machine with a new plate fitted and inked up with the appropriate colour, thus virtually, except that the same paper is used, repeating the work and the cost of the operation. Exact fitting or *register* of the two colours is essential, otherwise, for example, the blue line of a river will not fit correctly the V-shaped contours at the bottom of a valley, the black spot heights may not fit correctly between the brown contours, and a red road filling will appear outside the parallel black lines which should encase it. In deciding how many colours to use, another factor is the complexity and number of the features to be shown. While it will probably have been worthwhile (since it costs nothing extra) to have drawn these in different colours on the basic plot, the scientist may find after consulting professionals that they can in fact be shown in one colour without difficulty by using the right symbols in the fair drawing, both for lines and for areas. At this stage we are only considering hand-drawn or printed symbols and not the use of combinations of colour tints.

The second decision to be made is whether the map will be printed at the same scale as the basic plot and/or the fair drawing or at a reduced scale. The reasons for a reduction in scale are first that this may reduce either the size or number of sheets and thus the cost of printing them, and secondly that the reduction in size does a great deal to sharpen and improve the image by reducing the effect of imperfections such as minor irregularities in lines or letters in the fair drawing. This may save time in preparation or money in not paying a skilled cartographer to do it. In general, reduction involves photographing either, or occasionally both, the basic plot and the fair drawing in a *process camera*, while reproduction at the same size only involves contact printing in a *printing down frame* provided that the original is not on opaque material such as paper. Bearing in mind the size of the average map sheet it is clear that the camera must be a large and expensive piece of equipment and require more skilled management than the printing frame, and so here again the scientist must obtain estimates of the cost of the different processes before deciding this important question. He must weigh up these against the possible saving of a smaller map size (or number of sheets) and the advantages and possible savings of obtaining a better result from a less skilled fair-drawing stage.

Printing involves a combination of photographic and mechanical processes: photography to produce the images on the printing plates, and a machine to transfer these to a number of sheets of paper. The image is transferred to the plate by a variety of techniques which will be outlined later; all that the scientist needs to know here is that his fair drawing must have the maximum possible contrast between the image to be transferred and the background. Normally he would achieve this by drawing the image in black ink on a transparent and stable plastic material such as *Astrafoil* or *Permatrace*, but he should be aware of another technique called *scribing*, because although it requires special tools and training it can more easily produce a cleaner result especially for curved lines like contours or soil boundaries. In scribing the image is transparent and the rest of the material is photographically opaque; the scribe coat is covered by an orange dye through which the lines or symbols are cut with special tools to produce transparent lines. Whichever technique is used the important point is that the opaque areas must appear completely opaque to the sensitized surface of the printing plate, and the transparent areas must appear completely clean and clear. Ink drawings are in *positive* form; scribed drawings are in *negative* form.

As has been explained above, a separate printing plate is required for each basic colour in the final map, and following this backwards it is clear that a separate fair drawing is normally required for each colour in the final map (although exceptionally a multiple fair drawing may be separated out photographically). In the first case each fair drawing, or in the second case each separate negative made from a common drawing, shows only those lines, symbols, and lettering required in that colour. Thus for multicoloured maps the basic plot must not only be fair drawn but at some stage it must be *colour separated*, with the fair drawing for each colour normally prepared either in black on a clear background as a positive or cut out of an opaque background to form a negative. These must be traced accurately from the basic plot, and each must include *corner marks* to ensure that they fit exactly and that the images printed from the plates made from them also fit exactly on the paper. It will be clear from this that creating a multicoloured map not only multiplies printing costs but also requires more skill and material for making the fair drawings.

In summing up this introduction to cartographic and printing processes it is clear that before the scientist who plans to fair draw his own map puts pen to paper (or decides to learn how to scribe) he must make the following decisions about the final printed map:

(i) the number of colours to be shown on it;
(ii) the number and size of the map sheets required to cover the area;
(iii) whether it will be at the same scale as the basic plot or on a reduced scale;
(iv) whether to use ink drawing or scribing or a combination of the two.

In the rest of this chapter we describe in more detail how to prepare the basic plot for fair drawing, whether this is done by the scientist himself or others. The method of making the fair drawing is described in the next chapter, as are, in less detail, the following processes which the scientist is unlikely to attempt himself:

(i) *reproduction*, i.e. how the fair drawings are reproduced as printing plates;
(ii) *printing*, i.e. how the images on the plates are transferred to the final copies of the map on paper.

One important last point must be made. This is that a map showing only non-topographic features (such as soil or minor vegetation or land-use boundaries, geological or geomorphological features of specialized but not general interest, and so on) is by itself of very little value. The reason for this is that the user is unable to relate it to the ground without any easily identifiable and visible topographic features, such as houses, villages, paths or roads, streams, ridge lines, etc., by which he can know where he is. This combination of topographic and what we may call specialized scientific features can be achieved in three ways. In the first the specialized features are shown on an *overlay*, i.e. a specially printed transparency showing only the special features, which is laid over the topographic base map and is of course at the same scale and keyed to it. In the second the special features may be *overprinted* on to a special version of the base map which is printed in one subdued colour so that it does not obscure the printing of the special features but is available to relate them to topographic features where required. If taken from an existing map this is likely to be simplified in some way, for example by leaving out the contours and printing the black and the blue features (usually the manmade and water features) in the one subdued colour. In both these cases, where a good map already exists, the scientist will be using the kind of techniques outlined in Chapter 10. In the third case, where no adequate topographic map exists, the scientist will either have to survey the topographic features himself (on the ground or from air photographs) or will be working in conjunction with professional surveyors whose job it is to survey the topographic features and help him to relate the specialized features in which he is interested to them. Here also, in the maps for which the scientist is responsible, the topographic and non-topographic features may then have to be separated later before producing the final result in one of the two ways described above.

Mapping natural features

General

The three most important characteristics of the basic plot are that it should be clear, consistent, and definitive. Clarity means that all the features on it can be seen and easily distinguished from each other; consistency means that the same features are shown in the same way on all the map sheets (even when different workers have prepared them) and that their positions correspond at sheet edges; and being definitive means that no uncertainties are left about the identity and extent of the features shown or that if uncertainty is unavoidable this is clearly stated. All three requirements mean that before he starts detailed mapping the scientist must decide what features he is going to show, how they are defined, and how they will be depicted. There are existing models he can follow. In topographical surveys he should follow as far as possible the style of local maps at about the same scale or those produced by international or other aid agencies for that or similar areas. However, he should remember that published maps include symbols drawn by skilled cartographers and that in the field he will have no chance of using the specialized tools and printed symbols or letters accessible to them, and so must be content with simpler versions of these, supplemented for clarity by the use of different colours.

Both in the field and in mapping from air photographs in an office the scientist will normally use pencils initially because these will not wash out if the sheet gets wet and because mistakes or provisional delineations can be erased easily with a rubber. Although for very precise work hard pencils, up to 6H are used, the scientist will find that 2H or 3H is quite hard enough for the sort of precision with which he will be working; for provisionally sketching in features or for explanatory notes and sketches to be deleted later he should use an H or an HB pencil. Pencils tend to be softer in high temperatures and harder in low ones, so the grades to be used should be varied accordingly. Once part of a sheet has been mapped definitively in the field or office it should be inked in, normally in camp or at base, using inks recommended by the manufacturers of the base material. It is at this stage that different colours should be used, which are normally black, brown or red, and blue. However, if dyeline (diazo) prints are to be made for advance use from the basic plot, colours must be chosen which will reproduce by this process. Special inks for this can be obtained. A separate pen with a fine nib should be used for each colour, and drawing can be entirely freehand following the pencil line exactly; lines need not be of constant width and no special ruling pens or pens fitted with reservoirs need be used; these are described in the next chapter for use in fair drawing but are unsuitable for use in field or camp conditions. After inking in, the pencil work is erased. Black is used for the sheet margin and grid (if it has not already been drawn professionally), for most manmade features, and for all names except of water features; it can also be used for rock or gravel. Blue is used for all water features and brown for contours and sand or mud. Where non-topographic features are being mapped the scientist can vary the colours to suit his own convenience. To achieve consistency, especially with several workers on the same project, it is essential at an early stage to select and classify the features to be shown, decide on the way they will be shown, and prepare a *legend*, i.e. a descriptive list of all the symbols accompanied by examples of them in the correct colours. Each field or air photograph worker should have his own copy. This legend will then accompany each field sheet when it is passed for fair drawing, and before this begins agreement must be reached with the cartographers on how they will translate it into the symbols and colours (if more than one) to appear on the final map. Areas can be coloured by painting with water colours or shading with crayons.

In this way consistency of depiction will be achieved. However, consistency of position must also be achieved across sheet edges or where the work of two different scientists meets. All sheet edges must be compared before the sheets are sent for fair drawing, and any discrepancies in position must be removed. If the discrepancies are serious, as when a feature cuts off completely at a sheet edge or there is a major shift in position or size, the two scientists must confer and agree on a common solution, or a note must be added to the sheet edges. Minor cases can be resolved by taking a mean position across the sheet edge and adjusting the work on either side of this by eye. Another common fault is failure to complete boundaries of different types of area or to carry contours right through, and this must be checked on all sheets. All areas, e.g. of different soil, must be labelled. The best way to ensure this kind of consistency is for each field worker to submit his inked-up sheet to another worker for examination; if the 'examiner' is a scientist in a different speciality so much the better. If *he* cannot understand the sheet when it is accompanied by the legend, then it is unlikely that a non-scientific cartographer will be able to do so. In large projects the chief scientist should carry out a further check himself after this cross check between fellow workers.

Proper definition of the features to be shown often involves a more fundamental change of attitude by the scientist. He has learned to be cautious in making definitive statements since later research may disprove them, and it is quite understandable that in a difficult area he may be unable to say exactly where the boundary between two soil or vegetation types really lies or to define exactly the limits of some geological or geomorphological feature. This is of course how anomalies occur at sheet edges. However, the field or air survey worker must resist the temptation to leave this uncertainty to be resolved later (unless of course it depends on subsequent laboratory work), and in particular he must not leave it to be resolved by the cartographer. For better or worse he must in most cases make up his mind, for no-one else will be in a better position than he is to decide the question of how to select or where to depict the features concerned. In a relatively few cases he may feel that it would be wrong to be definite, but here again he must make a firm decision and indicate this by the appropriate change of symbol or descriptive note. The common convention is to show any linear features whose position is uncertain by *pecked* instead of solid lines, and instead of showing detailed shapes to draw smooth and rounded ones. Where the identity of a feature is in doubt, a question mark (?) can be added to the description or symbol. These items must also be decided on before work starts and shown in the legend. The principal natural features to be shown on a topographical map are outlined in the remainder of this section and some guidance is given on how to deal with them.

Natural features can be divided into the following groups:

(i) planimetric features such as coastlines, streams, lakes, marshes, mountain peaks and ridges, dunes and forests;
(ii) the shape of the ground, i.e. including the third dimension of height;

Streams, if not shown in blue, can often be distinguished from roads and other linear features by their irregularity. In air surveys they can usually be shown accurately; in the field the irregularities may have to be added conventionally to the stretches sketched in between fixed points. Stream beds which are normally dry can be shown by short pecks or dots, and where the scale of the map and the width of the stream bed allow they can be shown by a double line with a filling composed of lines parallel to the banks for a flowing stream and dots for a dry one. Where the stream is hidden by trees or is inaccessible on the ground it should be shown by longer pecks and smoother curves and 'unsurveyed' can be written along it. Coastlines, if well defined, should be shown by a firm line corresponding to the mean high-tide level. The mean low-tide level can also be shown if it differs appreciably from the high-tide level. The sand or mud between widely spaced tide levels should be dotted in brown, and rocky areas in black. At large scales, where the precise mean high- and low-tide levels cannot be defined exactly, the scientist should record the time and date when the water's edge was mapped and show any obvious high-tide marks such as lines of seaweed, driftwood, and other jetsam. In special coastal-zone surveys accurate contours must be mapped by levels or soundings, and these must be related to local tide levels by interpolation from existing tide tables or by a series of tide gauge readings taken at the time. Lake, reservoir, and river levels are usually subject to longer term, mainly seasonal, changes and in these cases the year, month, and day should be recorded against the waterline. Here also the highest levels can often be indicated by lines of weed, changes in vegetation, shingle terraces, or rock discoloration. The edges of marshes and swamps are often indefinite and seasonal or tidal; here no line should be shown but a gradual spacing out of the 'marsh' symbol. Where actual vegetation or ecological zones are being studied close contours related to tidal levels will be required.

Whether contours or less precise means are employed for the shape of the ground, the map must show the positions of the highest peaks and ridges since they define the watersheds between adjoining streams and with these form the planimetric skeleton between whose bones the actual slopes and shapes are fleshed in. The tops of steep cliffs are usually quite definite and can be shown by a hard line; the slopes below end less definitely and are indicated by a sketched rock symbol. The angle of rest of loose material is usually about $30°$, and cliffs can be defined as any slope steeper than this. In glaciated areas permanent snow or ice fields are best shown by a light blue wash or shading with contours, and any features such as crevasses, melt stream beds, or ice falls can be drawn in blue. Moraines and rock falls on the ice are shown by overlying black dots or stipple, but they may well obscure the exact edge of the ice. Permanent ice fields can usually be distinguished on air photographs taken in winter by their smoothness or by obvious ice falls or crevasses. In deserts the depiction of rocky areas and dry stream beds is straightforward but an attempt should be made in flatter areas to distinguish between those that are suitable for wheeled vehicles and those that are not; this is of course easy to do on the ground but rather difficult to determine from air photographs. Dunes show up well in air photographs; however, their types vary and the scientist depicting them should consult local maps and previous publications about the area. In detailed studies of dunes and glaciers the volume and exact shape may be important, and this can only be shown by close contouring and/or by a symbol depicting the dune or glacier crevasse pattern.

While specialists may wish to show a considerable variety of species in forest areas, topographic maps should concentrate on practical differences of interest to ordinary users such as the height, density, and coverage of the vegetation canopy, whether the trees are coniferous or deciduous in temperate zones, and whether or not they provide good cover and shade in the tropics. In savannah and semi-desert there are some species of large individual trees which are large, long lived, and widely spaced; these may be worth showing individually as landmarks. At very large scales both the trunks and the canopy outlines of each tree may be required. Tracks, paths, and firelines inside forests are important as lines of communication and should be mapped at any except the smallest scales.

The shape of the ground (relief)

As has already been explained (p. 185), the addition of contours to a map, especially one made from air photographs, adds a great deal to the cost of the survey. Nevertheless, contours are the only way of showing the shape of the ground with an accuracy comparable with that of the planimetry. Having decided that contours are required, a decision is required on the contour or *vertical interval* (VI). Relevant factors are the steepness of the slopes, the use to which the map will be put, the increased cost and effort of close contouring, and the proposed scale of the published map. In rugged areas too close an interval is not only expensive but may result in an illegible brown sheet of paper; doubling the interval would have been cheaper and more effective. In general contours should not be less than 2 mm apart even on the steepest slopes. If the area includes both steep and gentle slopes the interval may be varied to suit these with extra contours interpolated in the flatter areas and shown by a special chain line (. . . . .). Alternatively, if one type of area is irrelevant (as in maps of flood or outwash plains bordered by steep hills whose slopes are unimportant), these slopes can be shown by more approximate means. Contour values are traditionally written to be read looking up hill, and every fourth or fifth contour (or those corresponding to tens or hundreds) should be thickened. Checks must be made that the contours are in sympathy with the spot heights and trigonometric values. A simple rule of thumb in choosing the interval for undulating areas is that its value in feet should be much the same as the number of thousands in the scale ratio; for example, a 1/50 000 map should have a 50 ft (15 or 20 m) interval, and a 1/5000 scale map should have a 5 ft (2 m) interval. On any slope over $30°$ the contours would then be less than 1 mm apart and would not be shown. In very flat areas contours may be so far apart as to be almost meaningless, and the ground shape is better shown by a grid of spot heights.

Where the extra effort and cost of contouring is not justified, the shape of the ground, especially in rugged terrain, can be shown by one of three techniques: *hachures, form lines,* or *hill shading. Hachures* are lines drawn down the slope and vary in thickness and frequency with its steepness. They require great drawing skill to be effective, and are now seldom used except to show small areas of steep slopes such as terraces or cuttings on large-scale uncontoured plans like those of the British Ordnance Survey. They are unsuitable for showing general shapes on basic plots. *Form lines* are approximate contours, i.e. they are horizontal across the slope, and are normally shown pecked rather than solid; moreover, they have no values and are not continuous through the area since they show local shapes and not relative or absolute heights. Form lines are very suitable for field plots, but they can also be used on surveys from air photographs. It is essential to establish first the planimetric positions of the streams, the ridge lines, the peaks, and any breaks of slope between flat and hilly country; the form lines can then be interpolated between these. *Hill shading* is really a fair-drawing technique which is done with a soft black pencil by shading the south-eastern slopes to give the appearance of a relief model illuminated from the north-west. The density of the shading is varied to show both the steepness of each slope and its direction relative to this illumination. When well done it gives a very vivid picture of the shape of the country, especially when this is rugged, and it is often used on published maps to augment the more accurate information supplied by contours. It is unsuitable for field surveying, but can be applied quite easily when mapping from air photographs; in fact if these were taken with the sun in the north-west (which is rare in the northern hemisphere) the photographs themselves are already correctly shaded. This appearance can even be used by printing air or satellite photographs, correctly scaled, as a background in the reproduction stage. For compiling an air survey plot the photographs should be studied in stereoscopic pairs and the slopes shaded in by hand, the stream and ridge lines having previous been mapped to act as a framework. Hill shading is more difficult to erase than form lines; it requires more skill and it should be practised before work is started on the actual plot itself in order to ensure uniformity.

Mapping manmade features

Visible manmade features comprise communications, buildings, property, demarcated administrative boundaries, cultivated areas, and quarried areas. Railways can include not only main lines but light railways used on plantations or in quarries; a railway on the basic plot should be shown by a single black line with cross ticks, and the gauge and number of tracks should be noted so that at the fair-drawing stage it can be shown by the appropriate symbol. On air photographs railways usually appear to be narrower than roads and their curves are generally more gentle. For the map user the main concerns with roads are the type of construction and surface, and whether it is all-weather or seasonal; various symbols are used on published maps to depict these different characteristics. The basic plot should include information about these in note form, and all forms of stream crossing should be distinguished including bridges, culverts, Irish bridges (causeways), fords (vehicle or pedestrian), ramps for ferry terminals, and natural crossings of dry river beds since these indicate better than any other feature whether the road is only usable in dry weather. While air photographs show the shape of the road and can be used to judge the amount of engineering design, it is not possible to identify on them whether it has been finished, whether it has been maintained, and what sort of surface it now has; the classification of routes is normally a major part of the ground check or *field completion* of air survey maps. With natural surface tracks the important point is whether they can be traversed by ordinary cars, by four-wheel drive vehicles only, or not at all. In desert areas a track spreads widely over soft sand and only a general indication of the route can be given; in mountainous areas it is likely to be single and narrow and can be defined accurately. A path is best defined as a track too narrow for four-wheeled vehicles which is used by pedestrians, riders, or cyclists. On large-scale maps gates, stiles, and foot bridges should be shown. Normally roads and tracks can be shown by a single line unless the scale is large enough for the scaled width to be more than 1 mm.

Transmission lines can vary from a major feature with high-voltage wires carried on elaborate pylons from a dam or other power source to feeder lines carried on single poles to farms or small settlements. Individual pylons or poles should be shown unless they would be only a few millimetres apart. Transmission lines (and overhead cable ways) are difficult to see on air photographs except where lines have been cleared through forests; turns are sharp and the shadows or large pylons and the disturbed earth at their feet can be seen on photographs if the scale is large enough. In mapping buildings much depends on the scale and the importance and relative isolation of the building in relation to the map's purpose. At very large scales outlines of individual buildings are shown, including garages and huts, although decisions are required about their probable permanence. In the Caribbean, for example, 'chattel houses', although sometimes quite large, are made of wood and are removable. The small grass or mud huts and sometimes more complex structures of a developing country's rural settlements may require a very large scale for clarity and to allow room for names or numbers relating to occupancy or ownership, but this does not mean that such a plan need be plottably accurate to 0.5 mm. In professional large-scale surveys buildings are shown by their ground line and not their roof line, which is sometimes all that can be seen from air photographs, but normally the scientist need not bother about this difference. The interior of the building outline should be shaded or dotted; at fair drawing a *stipple* (a fine pattern of dots) or other device will be used. At scales smaller than about 1/5000 some generalization will be required, and to ensure uniformity the minimum size of small huts, and of narrow alleyways too small to be shown, must be decided at an early stage. Individual buildings may be important either because being isolated they are landmarks and shelters, or because in a village or town they have special functions such as police stations, courts, chiefs' houses, religious buildings, town halls, hospitals, or dispensaries. Often the functions of these are required and should be shown by symbols or names. It is important when generalizing built-up areas to retain the character of the area by maintaining the differences between building size and density.

Physical boundaries

Physical boundary features like fences, walls, hedges, or the edges of terraces or irrigation pans are relatively easy and straightforward to map, especially from air photographs. To proceed from this topographical mapping process to distinguishing those features which mark boundaries of ownership or occupation, whether legally or not, is seldom straightforward and may be quite impracticable, either because of the passions it arouses or because of the complexity of local land tenure systems. In the United Kingdom the large-scale plans (1/1250 in towns, 1/2500 in the country and 1/10 000 in moorland

and mountain areas) only show a single line for the sometimes multiple physical features which separate the land parcels without any commitment to whether these coincide with ownership boundaries: with a hedge and bank and a ditch it is the hedge which is shown. While they show features which appear to demarcate separate fields, factory areas, gardens, or yards, they omit those which appear to be minor and internal such as ornamental hedges or walls in a small garden. Although these plans are generally used as the basis for individual title plans in private conveyancing or in the government registration of land, the details of ownership are not available to the public. In other countries, where a similar comprehensive large-scale topographic plan exists, it may have no significance as a guide to ownership which is given by the cadastral plans described in Chapter 10; sometimes, because of errors in one map or the other, or because the later physical boundaries have not followed the earlier cadastral ones, there may be substantial differences.

It is clear from the foregoing that even in the United Kingdom the coincidence of visible physical boundaries and the legal property boundaries cannot be assumed, and overseas it can almost never be. This does not mean that the physical boundaries are not worth mapping; this may be relatively easy to do, particularly from air photographs, and once it is done the task of relating the ownership boundaries to these, although possibly requiring a great deal of investigation of documents and on the ground, may be a very simple survey technique requiring only the taping of short distances between the legal boundaries and clearly defined physical features already on the map. To carry out an elaborate detailed survey and an investigation of ownership or occupancy simultaneously is likely to be too large a task, and they are better separated.

Cultivated areas

Cultivated areas may range from large forest plantations down to orchards or walled gardens, and in deciding how many categories to distinguish the map compiler must be guided by three factors: the purpose of the map itself, the way in which it is being compiled, and the importance and likelihood of rapid change on the ground. Clearly, if it is a topographical map only easily distinguished and relatively few categories should be used such as forest plantations, orchards, open fields of cereals or grass and gardens; the temptation to show more and thus turn it into a detailed land-use map should be resisted. The main reason for this is that minor changes in land use, for example from grass to cereals, will lower its subsequent reliability as a topographical map on which the distinguishing feature of such fields was simply their openness. Moreover, although starting a more refined categorization is easy, keeping it up is much more difficult, and may involve the compiler in a great deal of extra effort which he had not foreseen. It may even prove impossible if, for example, much of the work is being done from air photographs on which green cereal crops cannot be distinguished from grass.

If a specific land-use survey is the object of the map then quite different factors are involved; in this case extreme detail may be required, and this may be specifically in order to be able to record later the changes which have subsequently occurred. The difficulty in this case, if the area is a large one, is to secure a reasonably instantaneous picture over the whole of it, even if the compilation of this into several map sheets is not so urgent. The technique for this will depend on the standard of literacy of the local population and, what is probably allied to this, the standard of the available topographical maps. Where both are high, as in the United Kingdom, then large numbers of local people, even school children, can be recruited, as they were in the first land-use survey of Great Britain, to mark up on the 1/25 000 maps (the smallest scale on which field boundaries are shown) the use to which each field was being put. Where this approach is not feasible, as in the land-use survey of Malawi some years ago, the only possible solution was aerial photography supported by ground checks of sample areas. In both cases the categories must be kept reasonably simple because semi-trained staff can, for example, distinguish cereals from grass but not wheat from barley, and orchards from plantations but not apple trees from pear trees if not carrying fruit, while the interpretation of different crop types on all but the largest-scale air photographs (which are often too expensive both to take and to handle) requires very great skill and special equipment. Sample surveys carried out on the ground over a relatively small area are of course fairly straightforward once the number of categories to be distinguished has been settled. Having worked out the number of categories to be shown, the simplest possible set of symbols should be used in the field. Where areas are large enough the category may simply be written across each; if they are too small for this a system of numbers or letters may be used, relating to a master list. Unless the survey has some unusual purpose it will be best to follow local usage (especially in a topographic map) or international usage in some other forms of survey. Since in most cases the categories will fall into major groups (for examples trees and crops) colour washes may conveniently be used to distinguish these on the basic plots, with letters or symbols added to distinguish the sub-categories. It is beyond the scope of this book to suggest the symbols or lettering to be used in more detail since each branch of science has its own conventions for these (Coleman and Shaw 1980).

Place names and administrative boundaries

The selection, allocation, and spelling of place names can be one of the most difficult tasks facing anyone making a map, especially if he is mainly using air photographs and does not spend much time on the ground or know the local language. The aim is simple: to allot to each significant feature a name which is unique and acceptable to both the local people and their government. Selecting the features is relatively easy, but allocation of names has two difficulties: correct choice of the feature and of the name. Both require knowledge of local customs and language and of how the environment is viewed. For example the apparent name of a mountain may only refer to a particular slope; similar slopes on other sides may have different names. An apparent river name may refer only to a particular stretch. Obvious traps like 'I don't know' or 'Just a mountain' must be avoided. Local aboriginal names may have been converted to words in a previous colonists' or invaders' language; does one then use the correct name or the later version?

Spelling of names is particularly complicated where they have been transliterated from another alphabet such as Arabic, Urdu, or Cyrillic into the 'Roman' alphabet used throughout the 'Western' world because several different forms of transliteration exist. This has happened because there is no exact equivalent in each alphabet of the letters of the other and because different Western nations pronounce the letters of the Roman alphabet in different ways. For example, the Arabic name for a dry stream bed is variously spelt wadi, ouadi, and uadi on English, French and Italian maps, while that for a mountain may be spelt jebel, djebel, or gebel. Some forms of transliteration, or of the direct writing (often by missionaries or anthropologists) of a local language may include a number of *diacritical marks* such as dots or cedillas written above or below letters, or even specially invented new letters for special sounds like clicks or glottal stops† for which

† As in 'an ice house' but absent from 'a nice house'.

there is no equivalent in the transcriber's language. These should be avoided if possible because while it is easy to include them on maps they will tend not to be used in typewritten or printed official lists and thus the name will have two different forms in current use. Special uses of standard letters like X or Q without U are more sensible and will not have this disadvantage.

A few simple rules are given below for the field scientist compiling a new map and wishing to add names.

(i) Consult existing maps and official government publications such as railway and bus timetables, and lists of post offices, dispensaries, missions, and farms, and look out for names on railway stations and signposts, and at the entrances to farms, mission stations, and other large establishments. Where discrepancies occur they should be discussed with their representatives. If a national geographical names committee exists it should be consulted.

(ii) If the scientist does not know the local language (or the regional *lingua franca* such as Swahili, Hausa, or Arabic) he should obtain the services of an interpreter, who should be literate and intelligent, to help him record the names in the most generally used language.

(iii) In remote areas, where both a general and a tribal language are used, names may be in the latter. In such cases local administrators, or missionaries who may have pioneered the writing of the local language, should be consulted as much as possible. A tape recorder will be useful in recording the spoken version for them to interpret, but beware of the fanatical expert who has invented a completely new form of transcription to replace existing ones!

(iv) Avoid using special letters or diacritical marks as much as possible, but accept standard letters used for special sounds like clicks or glottal stops.

(v) Compile a glossary, with aids to pronunciation, of the commoner feature names in the area, and add this with translations to the map, giving where necessary both the local and regional language equivalents.

(vi) In uninhabited areas, few of which have now not been visited before, consult all the records of earlier expeditions and also the government which exercises jurisdiction over the area before allotting new names.

(vii) Consult the bodies with long experience in such areas: in the UK these are the Permanent Committee for Geographical Names, the Scott Polar Research Institute, and the Foreign Office, which has responsibility for names in the British Antarctic sector.

Administrative boundaries

Administrative boundaries may be international or internal. The former, like the disputed names of international features, can cause a great deal of trouble and are best omitted from maps if there is any danger of a dispute. If they are shown the map should carry a footnote stating that it is not to be regarded as an authority for boundaries. Internal boundaries are normally less controversial even though many of them have not been precisely demarcated on the ground or mapped. Where they follow streams, or water sheds, or in straight lines between well-known features such as mountain peaks, they should cause no difficulty. The main problems arise when the original definitions were imprecise or even impracticable (like the classic case of a turning point described as 'the place where we saw the giraffe'!). Another problem arises when a combination of named topographical features and approximately surveyed locations or bearings has been used and a later more accurate survey shows that these do not fit together and that there is an ambiguity. As usual the visiting scientist should consult local opinion at all levels and 'keep his ear to the ground' because even internal boundaries between tribes or villages may be hotly disputed and both sides may regard him as some kind of government representative or impartial observer who may be influenced into making a judgement in their favour which can then be used to support their case. Ideally a boundary should be settled by agreement between the authorities responsible on both sides of it, then demarcated with permanent monuments and finally surveyed so that any attempt to move the demarcating monuments can be detected; however, this ideal is rarely achieved in practice. The decision on whether to add such a boundary to the map at all should depend first on whether it is required for the purposes of the map, and secondly on whether it is clearly enough defined, and free from dispute, to be shown without causing controversy. In England and Wales public boundaries such as those between parishes have been 'mered' or defined on the ground by an Ordnance Surveyor accompanied by representatives from both sides and full records are kept at the Ordnance Survey in Southampton.

A particular case of difficulty arises where a boundary, whether international or internal, lies under water as in an estuary or river or between two islands, particularly now that the sea bed has become potentially valuable. For obvious reasons it is rare for such boundaries to be demarcated and they are usually defined either as the median line equidistant from both shores or banks or the *thalweg* of the river, i.e. the lowest line of the river course, but these immediately give rise to uncertainty and dispute because, for example, the methods of measuring the distances from the shores to the median line are open to different interpretations. For example, one could take high, low, or mean tide lines as defining the sea shore and the banks of a river as being at high or low level; and clearly it is impossible to measure the distances from every point along each shore so some generalization, like drawing straight lines between headlands, is necessary, and different forms of this will produce different results. The ownership of off-shore islands in a strait between two countries (or provinces of a country) may also be in dispute, and in seasonal rivers islands of valuable agricultural land which only appear in the dry season will amost certainly be gradually changing shape or even disappearing and reappearing in a different situation. Then the question arises as to whether it is the same or a different island altogether. One answer which has been used on the Nile has been to 'settle' or 'adjudicate' the boundary between the villages on its two banks along a stretch of river once and for all by reference to permanent monuments above high river on both banks and to produce a map of it; only those parts of the boundary above water at low level at that time are then demarcated by a surveyor, and the two sides accept that over several generations they will expect to have ownership of equal areas of flood plain and they will engage in temporary leasehold agreements for its use as the river favours temporarily one side or the other. The scientist should try and avoid becoming involved in boundaries of this type and should not show them on his map unless specially required to do so. Many map agencies include a small boundary box in the footnotes which gives only their approximate locations (see p. 296).

12. FAIR DRAWING, REPRODUCTION, AND PRINTING

Fair drawing and scribing

General

Several books are available describing in great detail the materials, equipment, and techniques used in fair drawing; the best known to the author are listed in the Bibliography. Only an outline of the subject is given in this chapter since any scientist interested in preparing his own fair drawings should seek advice from, and if possible practise with, a professional cartographer; for example, some geography courses include elementary training in fair drawing. No textbook can be a substitute for training and practice in a craft where much of the skill depends on aptitude. In deciding whether to attempt his own fair drawing rather than leaving it to a professional cartographer the scientist should consider first whether he has this aptitude, which requires more the qualities of manipulative skill, neatness, and patience, as well as some artistic feeling for pen and pencil, than a scientific intellect. If he lacks these qualities he should probably not attempt it. The other important factor in deciding whether to do it is the time he has available from his main tasks. He should remember that the average cartographer is less highly paid than himself even as regards salary, and when the full cost, particularly overseas, of employing a field scientist, including accommodation, transport, and local labour, is considered it is very uneconomical for him to spend time in cartography if it is at the expense of doing his proper job. However, there may be occasions when, for example, he is confined to the office or tent by bad weather or by having to supervise some long-term experiments, and fair drawing his field sheets may be a suitable way of using the time; he may even have time to find and train local assistants. It must be remembered that, as with other mapping skills, to acquire something of them is also to acquire a better understanding of the problems and motives of professional cartographers with whom he may have to deal on another occasion, particularly in some large project in which all the work must be divided up among the specialists – including the cartographers – best qualified to do each part of it.

Materials

A variety of drawing materials such as paper, tracing linen, plastic drafting medium, pencil, biro, and inks are available for the basic plot, but the requirements for fair drawing are more specific if the result is to look anything like a professional job. The material used for the basic plot should if possible be dimensionally stable and transparent, and so plastic drawing materials are recommended for this; but they are essential for fair drawing. Distorted basic plots can be copied square by square and opaque ones can be traced, but distorted fair drawings are useless (especially where more than one colour is to be printed) and opaque ones will require the use of a process camera before printing plates can be made from them; fortunately suitable materials other than glass are now available in the form of polyester or polyvinyl materials. These can be obtained with one or both sides *grained*, i.e. with a matt surface which takes ink, and they are essential for map sheets of any appreciable size. Polyesters such as Permatrace are preferable; they do not shatter if dropped on an edge as polyvinyls can, and they are dimensionally a little more stable. However, polyvinyls such as Astrafoil can be used but most be treated with more care. Sheets 0.01 in (0.25 mm) thick which cannot be rolled can be used for the most advanced work, but sheets half this thickness cut from a roll are quite adequate for many purposes and are of course cheaper. Material which is smooth on both sides should not be used. Paper must not be used for the reasons given above.

Ordinary waterproof ink should not be used for fair drawing because it flakes off plastic; the maker of the sheet material should be consulted on what inks to use. As already explained, only one colour, black, is used in order to achieve the maximum contrast between the drawn image and the background. To test this opaqueness the sheet should be viewed against the light from the reverse side. Where the fair drawing is to be scribed and not penned the scribecoat surface is covered with a special emulsion coating over a polyester or polyvinyl base out of which the images are cut with special tools. The material is known as either a *positive* or a *negative scribecoat*. Positive material has an opaque white background, and after the lines have been cut out of this a black material or dye can be applied so that the lines appear black as if they had been penned in ink. Negative material has an orange dye over it which is translucent so that the plot being traced can be seen through it, but it is opaque to the light used in

284 FAIR DRAWING, REPRODUCTION, AND PRINTING

Fig. 12.1 Scribing.

the photographic process. The result will be a negative and the image must therefore be the wrong way round, so it must be traced from the *back* of the plot and not the front. Tracing can be done by putting the plot and the fair drawing material over a *light table* (see below), or, if the plot is on opaque material, by following a mirror image photographically produced in blue on the scribing surface. Clearly the latter is only possible by using a process camera, either in a professional cartographic establishment or overseas if the local survey department is equipped for this and prepared to do this work for the scientist in some sort of joint project. He will need to have small quantities of the scribecoat dye so that errors in the cutting can be painted over and allowed to dry; a fresh correct image can then be cut out of this.

Equipment

Although fair drawing can be done on an ordinary table or even if necessary in camp on a plane table or drawing board, for any extensive programme it will always be worthwhile to buy, borrow, or make a *light table* (Fig. 12.2). This consists of a hinged frame at least 1 m square which can be set at a convenient slope and height for working. The middle of the frame is fitted with a rectangle of thick glass which

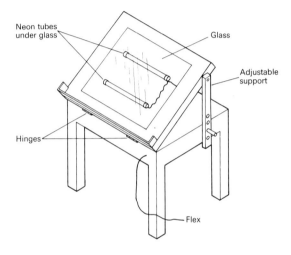

Fig. 12.2 Light table.

if possible is as large as the map sheets to be drawn. It should have its own legs and support and not be designed to fit on an existing table; scientists will find that once they set up a drawing office there is never any table space to spare. The light table can either be set up opposite a low window, so that daylight shines through it from underneath or, better, if electric power is available two neon tubes should be fitted a few inches below the glass in a box so that they give a strong but cool light from below. Since most work nowadays is done on transparent plastics, this makes it easy to trace off from a plot underneath and it is even possible to do this with paper if it is not too thick; a light table is essential for scribing. There should be a small ledge at the bottom of the frame to hold pencils etc. and an attachment at the left-hand side for a table lamp.

Since a map consists mainly of lines, the most important difference between a basic plot and a fair-drawn map is that in the map all the lines have a constant width except those representing streams of varying width. This cannot be achieved with an ordinary pen, and three main types of tool have to be used. The first is the *ruling pen* and its variations (Fig. 12.3(a)). It consists of two steel blades of equal length with slightly rounded ends (i.e. not sharp points) set in a handle and hinged at the upper end; the separation of the lower tips is adjusted by a fine screw against a spring. The ink is inserted between the blades and these define very sharply the edges of the line whose thickness is equal to the separation of the tips of the blades. For straight lines the blades are fixed directly to the handle, but for contouring the blades are curved and fixed to the handle by a swivel so that they can follow the curves of the contour. The scientist will have to learn the technique of shaping and keeping the ends of the blades sharp and correctly rounded, and they must always be cleaned before inserting a fresh supply of ink. Drawing must be done against some sort of guide, and it is very much slower than drawing freehand, but the lines will be of constant width and free from unsightly irregularities. The guide can be any straight piece of plastic or wood such as a ruler, a scale, or a set square; *French curves* (Fig. 12.3(b)), which have varying curvatures to fit the curves of contours etc. to be followed, or splines held down by weights (Fig. 12.3(c)) can also be used. For drawing double lines to represent roads or railways, a road or railway pen (Fig. 12.3(a)) is used; in this not only the pairs of blades are adjustable but also the distance between them which represents the width of the road or railway symbol. The scientist will be familiar from school days with the ordinary pair of compasses fitted with a pencil or lead; for drawing circles in ink he must have a ruling type pen fitted instead of these. Examples are shown in Fig. 12.3. All these pens have to be filled frequently using a quill or ordinary pen, and work with them is very slow; modern types of pen have in general overtaken them. These include special nibs which are pre-set to special gauges to draw lines of specific width, and they are fitted with reservoirs which give a constant flow of ink. The *Graphos* is a common type (Fig. 12.4); however, it may prove more difficult to maintain in camp conditions, and if locally trained cartographers are employed they may be used to the old-fashioned types of pen. Plastic reservoirs may only be able to take certain types of ink. No skill is required for re-shaping the nibs which are thrown away when worn or blunted.

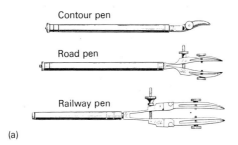

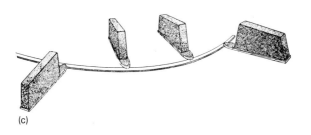

Fig. 12.3 Drawing pens and curves: (a) contour, road, and railway pens; (b) French curves; (c) spline and spline weights.

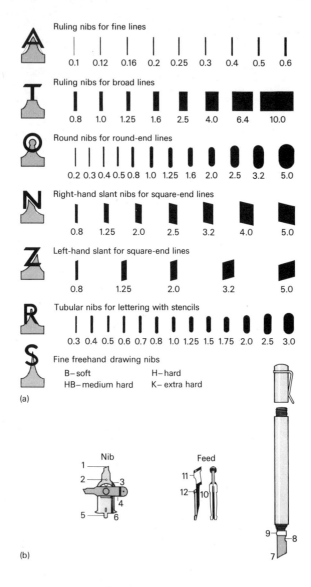

Fig. 12.4 Graphos pens and nibs: (a) nibs; (b) pens: 1, nib slit; 2, ink passage; 3, nib hook; 4, pivoting cover; 5 nib tongue; 6, protruding end; 7, feed aperture; 8, filling opening; 9, slot for nib tongue; 10, feed duct; 11, ejector; 12, filling grooves. There are three types of ink feed: no. 1, flowing sparingly; no. 2, medium flow; no. 3, freely flowing.

Scribing

As already mentioned in Chapter 11 the decision whether to use penning or scribing techniques is a major one which must not be taken in a hurry or without adequate preparation. The scribing tools and materials are more expensive than those for pen and ink drawing, but the scribing technique is often more easily learnt than that of using a ruling pen. Scribing is particularly good for drawing curved lines such as contours or soil or vegetation boundaries, while a ruling pen and straight edge are better for drawing rectangular features such as houses. Scribing is worth considering if the job is a large one, and if the scientist has the time to learn how to do it and to train assistants. Once mastered, the technique has considerable advantages in the greater speed of working and in the sharpness of the finished result. Another advantage is that the result is normally a negative (i.e. clear on an opaque background) from which a printing plate can be made directly, thus in most cases saving an extra process. No written guide can be a substitute for practice under professional guidance, so only a brief description will be given here. The equipment consists of special cutting tools. These include a small platform standing on two ball-bearing feet with the cutting tool in a special holder at the front. This swivels in a ball race so that the cutting or chisel edge is always at right angles to the direction of drawing, as with the swivelling contour pen described on p. 286. For any except the thinnest lines various widths of chisel point are supplied, and if correctly handled each will cut exactly the required width of line with sharp clean edges. For very narrow lines a point is used with a fixed holder. The cutting points or chisels are made of synthetic sapphire which is initially expensive but has a long working life. They do not have to be sharpened but can chip and be ruined if they are not handled with care. As with ruling pens, double points are available for drawing the parallel lines required for roads and railways. The chisel points have a 'flat' so that they can only be inserted the correct way round in the holder and are held rigid by a small grub screw. For scribing small dots a chisel point is supplied to be held in a special holder in which it can be rotated to produce a perfect circle. Tungsten cutting points are also available and are cheaper than sapphire, but they require sharpening which is a skilled job. It is clear that great care is required to ensure that the cutting tool operates at the correct depth in order to remove all the coloured dye without cutting into the material beneath (Fig. 12.1).

The elements of colour separation have already been described on p. 266. Where two or more colours are planned for the printed map the scientist must decide exactly which features will appear in each colour. Obviously they must be clearly distinguished on the basic plot, either by the type of line or symbol used or by colour. A separate fair-drawing sheet is prepared for each colour and the features concerned are traced on to it in black or scribed (Fig. 12.5). Each sheet can be placed over the basic plot and the relevant features traced off in turn, or a blue image of the plot can be printed photographically onto each of the fair-drawing sheets; these are then used for positioning while the basic plot is used only for clarification in cases of doubt. Where two line features of different colours run very close together they must be separated by small sideways movements so that they do not clash, but where they cross at a substantial angle this is not necessary. A light table is essential for checking against this type of clash, and also to ensure that all the drawing is completely opaque and not 'grey' where ink has been used and that in scribing the lines are clear from loose material or dirt. Exact register is essential, and this is best achieved by a series of punched holes and studs along the top margin so that each sheet can be lifted to check or alter what is underneath it. Corner marks must also be added, with ticks at right angles about 3 mm long. These will appear in all the colours on the finished map (although they may be hidden under the grid or graticule corners of the map sheet) so that the printer can position the plates to achieve exact register of the colours on the paper. They can be seen, if you look carefully, on any multicoloured map.

Lettering and printed symbols

It is now many decades since cartographers wrote all the names on their maps by hand, and the use of pre-printed letters and symbols is almost universal. Hand lettering is reserved for a few awkward cases where names have to be inserted into a small space, or spaced and curved to follow a stream or boundary, although even here it may be quicker and easier for the unskilled to cut out and stick down individual printed letters. Large cartographic organizations such as local survey departments (or local printing firms) will often have their own *letterpress* section which prints the names etc. from metal type on to paper and then photographs these onto film, but most organizations have modern photo-type-setting machines which produce

FAIR DRAWING, REPRODUCTION, AND PRINTING 291

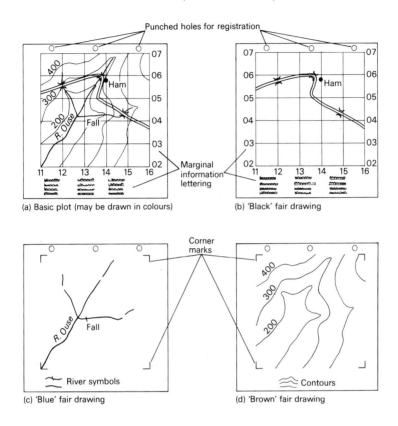

Fig. 12.5 Colour separation; (a) basic plot (may be drawn in colour); (b) 'black' fair drawing; (c) 'blue' fair drawing; (d) 'brown' fair drawing. Each fair drawing is drawn in *black*.

the lettering directly onto film to a size range of approximately 2-20 mm in height and in a variety of styles or founts of type. If possible the scientist should arrange in advance to obtain his type from such an organization, because he will then be able to have the complete names etc. printed exactly as he wants them; this will save a great deal of time in cutting them out and sticking them to his fair drawings. The disadvantage is that he cannot order them until each basic plot is complete with all its names and any legend or explanatory

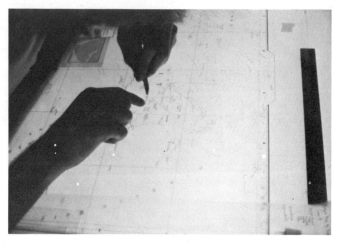

Fig. 12.6 Name sticking.

text he may wish to add to it, and he will then have to wait for his lists to be printed; if a local survey department is being used this work is likely to have a lower priority than their own internal requirements. However, if he is lucky and has planned well ahead, the printed names etc. should be available about the time that he has finished the line drawing.

Where no such printing to order is possible he will have to buy sheets of letters and symbols from an art material supplier or a stationer and on an overseas project this may have to be done in his home country or by post. He will then have to make up all his names and text by transferring individual letters. This is slow and tedious and requires almost as much skill as lettering by hand, but unless he has naturally neat and attractive handwriting the result will be very much better than handwritten names. A well-known form of this dry transfer pre-printed lettering is called Letraset, but other types are also available. In this case he will not have to know and list the names in advance, but he will have to make an estimate of the amount of lettering required in each style and size, and he should allow plenty of extra for mistakes, under-estimates, etc. The type should be on film unless the fair drawings are to be copied in a process camera, when they can be either on film or white paper; in a very remote and under-developed area the latter may be all that is available from a local printer.

The manufacturers of dry transfer lettering and of photo-typesetting machines have samples of different type styles and sizes available, and the scientist must choose what he wants from these, allowing of course for reduction if the fair drawings are to be copied at a reduced scale in a process camera. Various founts of type are available but he should choose one which is as simple as possible and without *hair lines* or *serifs* since there are liable to reproduce incompletely causing broken letters. Names can be in capitals (*upper case*) or in small letters (*lower case*), and sizes are defined by *points* which are a measure of height of the letters. In the USA and the UK a point is 0.0138 in (0.345 mm) but in Continental Europe it is 0.0148 in (0.37 mm); this page for example is printed in 10 point. Type size can also be quoted as the height of the capital E in millimetres. *Italic* lettering is sloping and is often used for water features, while upright lettering is used for the names of mountains, villages, and other human features. Figures will be required for grid or graticule values in the margins, spot heights, and contour values. Lettering may also be required for soil or vegetation types, for the legend describing the symbols in the margin, for explanatory notes, and so on. In choosing styles and sizes the scientist should be guided by existing maps that he has already followed for his line work and symbols. He must make out clear lists preferably typed or at least written in block capitals, with all the different categories clearly distinguished so that the printer producing the lettering, or the cartographer choosing from pre-printed stock, knows exactly what to do.

The positioning, alignment, and spacing of names are important, but because they do not have to be precisely located like the features themselves the map maker has a choice of where to put them for the maximum clarity and effect, provided of course that there is no doubt about the feature referred to. Since a name or letter once stuck down is difficult or impossible to remove, their positions should be planned beforehand and marked lightly in blue pencil (which will not photograph) either on the basic plot or on the fair drawings themselves by drawing a line on which the letters will stand and showing the length of the name. Heavy and black lines can if necessary be broken to give a space for a name, but very thin ones or those in lighter colours which a black name will cover do not have to be. Most names of small features and numbers or letters used for spot heights, soil types etc. should be exactly parallel to the top and bottom edges of the map; names of linear features like streams can follow them obliquely but should never be upside down; however, contour values should always read up hill and therefore can be upside down on the north side of a hill but this should be done as rarely as possible. Names of administrative or tribal areas should be in large but thin lettering and be spaced across the maximum dimension of the area. They are continued where they cross a sheet edge and the rest of the name is printed in small letters closely spaced in the margin, as shown below:

S U S S | ex and Suss | E X.

Many other conventions will be noted after careful examination of the maps being used as models.

Symbols

Map agencies and the firms supplying pre-printed sheets usually have samples of point and area symbols which can save a great deal of drawing and improve the appearance of the map. Point symbols include those for villages, houses, or important buildings on medium-scale maps, and such items as electric pylons, lighthouses, buoys, road or railway signs, etc. Area symbols include various types of *stipples* (patterns of dots), *rulings* (parallel lines), *cross hatching* (two or more rulings at an angle), and more elaborate patterns such as trees of different kinds, grass or marsh symbols, and so on. The scientist should consult the catalogue or sample cards in conjunction with his own ideas for the different areas he wishes to distinguish and classify. Area symbols are available on quite large sheets of film from which the required area can be cut. All lettering and symbols on film are normally supplied with a wax adhesive and attached to a backing sheet; this is peeled off and the piece of film bearing the lettering is then stuck down using a rounded bone or wooden handle. With large sheets of area symbols care must be taken to eliminate all the air bubbles under the film.

Marginal information

This is a suitable place to describe the margins of a map since these contain matter which is mainly explanatory and therefore requires a good deal of lettering. Most of this has already been referred to incidentally, but it is as well to bring it all together in one place for ease of reference. Which margin is used for the bulk of it depends on the design of the map and in particular whether, and in what way, it is to be folded. Although large maps are printed flat, and are stored in libraries and offices and stuck on office walls in this form, if they are to be used in the field or included in a report they will have to be folded, and it is essential to work out how this should be done at an early stage in the design of the map and its placing on the sheets of paper on which it will be printed. The map itself is surrounded by what is called the *neat line*, and everything outside this is the margins. In flat maps most of the information tends to be put into the bottom margin while the top one is relative narrow, but if the map is to be folded then the right-hand and left-hand margins may be quite wide and be designed to contain most of the marginal information. The scientist should consider this part of the design with care, take professional advice if possible, or at least study other published maps, and prepare a *mock-up*, i.e. an outline of the map and its marginal information on a sheet of paper the same size as that to be used for printing it. Where professional cartographers are being employed to fair draw the map, he should work out with them in detail how and where he wants this information to be shown and have them prepare mock-ups before they go too far with the fair drawing. In multicolour maps the fair drawings of the colours will of course have to include in the margins the relatively small areas of colour required, e.g. to show blue streams, brown contours, or special names or symbols in the legend. Most of the lettering and the neat line will be in black.

The principal items of marginal information are as follows:

(i) *Grid and graticule values* for the grid lines and the parallels of latitude and meridians of longitude. These will appear in all four margins, and they may be in colour although ticks or lines should be in black.

(ii) *A grid reference box* describing in detail how to give a grid reference.

(iii) *A scale* showing both the *representative fraction* (e.g. 1/5000) and lines with distances in suitable units of length at map scale: miles or kilometres for medium- or small-scale maps, and metres, feet, or yards for large-scale maps.

(iv) A *north point* in simple maps, and a diagram showing the *grid convergence*, i.e. the relationship between *true north* and grid north in more sophisticated maps; the direction of *magnetic north* should also be shown with the date and the amount of *annual change*.

(v) A *compilation diagram* showing the sources on which the map is based, unless it is all original work when a simple statement will suffice, e.g. 'based on field surveys by . . . during (year)' or 'compiled from air photographs taken by . . . on (date)'. Compilation diagrams are particularly important in derived maps where sources of different age and reliability have been used.

(vi) A *boundary box* showing on a small scale the main areas or countries covered by the map and the approximate boundaries separating them.

(vii) A note in large type giving the contour interval or vertical interval, though this should also be indicated in the legend where samples of the contours and other topographical re-representations are shown.

(viii) Any *disclaimers*, i.e. notes disclaiming responsibility for such legal decisions as rights of way, ownership, or administrative or international boundaries.

(ix) A *legend*, which is the most important and usually the most extensive part of the marginal information since it lists all the features shown with their symbols and in their correct colours. Sizes and types of lettering for names with their significance should be included, and also any abbreviations, e.g. PH for public house, PO for post office, etc. In a special scientific map showing complex areas and boundaries of different rock or soil types or species of vegetation this may be a very large and complex item; in fact in extreme cases it may have to be published as a separate sheet to be used in conjunction with each map sheet of a map series.

(x) A *glossary* of terms, if necessary with their equivalents in one or more languages used in the area and a guide to their pronunciation.

(xi) A *sheet index* showing the relationship of this sheet to the others round it and possibly the layout of the whole series.

(xii) The name and/or the number of the sheet, usually placed in the north-east or south-west corners in small type (for easy location in a stack of maps) as well as in larger type in the middle of the top margin. It is worthwhile selecting a name for each sheet of a series as a check on requests for it, since people are less likely to make a mistake than with a number alone.

(xiii) References to towns, villages, or other landmarks are given in the margin, with distances, where a road or railway or track crosses it. The remainder of area names which cross the sheet line are also placed in the margin as explained on p. 294. The names and numbers of adjoining sheets may also be shown.

Where the map is designed to be folded it is a good idea to put the name, and possibly a small sheet index, on the *outside* of the folded map, although this may mean printing on the back. If it is designed to have a cover then it is likely that this will be a more elaborate affair and will be specially designed and drawn to be attractive as well as to protect the map from wear, and will be printed separately.

Photographic reproduction

Field scientists engaged in surveying have already learnt that the word 'plot' means a survey drawing and not a conspiracy; in cartography they will find that *reproduction* refers only to photographic and printing techniques and not to biological processes. They will also find confusing the use of the word 'copy' to mean the original, especially when it is being reproduced by means of a process camera. Although printing plates can be made from positives (i.e. opaque images on a transparent background), it is more common to make them from a negative. Scribing will reproduce directly, but normally ink fair drawings have to be photographically reproduced as negatives before they can be transferred to the printing plates. There are two main methods of doing this:

(i) *contact printing*, in which the original drawing and the sensitive reproduction material are put together, face to face, in a *printing down frame* and exposed to light;
(ii) *process photography*, in which the original is placed on a *copy board* and photographed through a large camera.

In both techniques, because being face to face must mean a reversal of the image from right reading to wrong reading (or the other way round) and because all cameras cause a reversal of the image, the result will be a mirror reversal of the original. It is possible to obtain a reproduction the same way round by contact printing through the thickness of either the original or the reproduction material, but

298 FAIR DRAWING, REPRODUCTION, AND PRINTING

obviously the result will never be as sharp. This technique is only used for such purposes as producing temporary copies such as *dye lines* which are usually on thin paper and only cost a few pence each; they are often seriously distorted and dimensionally unstable. To ensure a really sharp result the image on the original and the emulsion on the reproduction material must be in contact, and for large sheets this requires that the printing down frame has a *vacuum back* which uses the air pressure in front to press the two firmly together against the glass, or, on a copy board, holds the original tight and flat against the board. Lighting in both cases is usually by carbon arc or other forms of high intensity lamps, and the exposure is judged either by experience or by making a few trial shots with a small piece of material. The result will be either right or wrong reading but it can also be either positive or negative.

Fig. 12.7 A process camera.

The use of a process camera must always add considerably to the cost of reproduction because it is a large and expensive piece of equipment and it is obviously more difficult to operate than a printing down frame. The camera may be simply an enlarged version of the old extending hand cameras, which are now seldom seen, operated in one room, or it may be installed in two rooms. It consists of three parts: the camera back which holds the reproduction film (glass is seldom used now) and is often built into the wall separating the two rooms, the bellows and lens unit, and the copy board which holds the original (known in the trade confusingly as the *copy*). The lens and copy board run on rails so that they can be moved relative to each other for focusing and for reduction or enlargement of the image. The rails ensure that the planes of the copy board and the film holder remain parallel and at right angles to the optical axis of the lens. Thus the image on the film will be sharp and exactly the right size and shape. Normally the fair draughtsman provides a template on transparent plastic showing the corner marks to which the negative must fit; this is checked together with the focus on the focusing screen at the back of the camera using a magnifying glass at each corner in turn before the focusing screen is removed and the film negative is inserted. In a modern camera the necessary adjustments can be made from the negative end by switches on a console which actuate electric motors, but in older cameras they may have to be made by hand. Where two rooms are used, that containing the copy board will also contain the lights illuminating this, and the other will be fitted up as dark room, with the usual sinks and drying racks. In addition to copying fair drawings the camera can also be used for reducing or enlarging basic plots to produce an image at a size convenient for fair drawing or for any task involving the accurate reproduction of one sheet of material at a different size; but the cost of doing this with the camera must be weighed against the requirement for accuracy since a result which may be good enough for some purposes may be cheaper to do by hand if the time and labour are available.

Whichever way the negatives (or positives) are produced, directly by scribing, or by contact, or by process camera photography, if more than one colour is involved it will be worthwhile making a *negative proof* from them. This is done by printing down each negative (or, with some systems, each positive) in turn onto a prepared white plate or sheet of plastic by an additive process which reproduces in different dye colours the image from each negative in turn. The colours will only approximate to those to be used in the final map, since they are dyes and not printing inks, but the negative proof does fulfill three important functions without incurring the cost of making separate printing plates, setting them in a proving press, and running off paper copies. However, since the production of a second negative proof involves repeating the whole process, normally only one copy is made, and if anyone other than the cartographers or photoprocess workers in the organization wish to see it they will usually have to come there to do so. It fulfills the following three functions.

(i) It enables the draughtsman, and possibly the scientist if available, to see for the first time (except for the actual colours) what his final map will look like; up to now it has only existed as a basic plot or in his imagination.

(ii) It will expose any clashes in the different colour images which have been overlooked in the earlier comparisons on the light table.

(iii) It will disclose to the scientist or cartographer any errors or inconsistencies of fact or style which were not apparent in the earlier rougher versions, in the same way that a fair-typed version or a printed proof of an article or book reveal inaccuracies and inconsistencies to the author.

A negative proof is therefore well worthwhile in any complicated multi-coloured map, and the scientist should if possible see it. This is particularly true of the first sheet of a new map where the style may still have to be finalized. However, this must be thought out ahead so that this sheet is completed well in advance of the others (or it may even be a mock up which is not based on a real basic plot but includes all the features which are to be shown), otherwise any changes of style decided after seeing it will involve redrawing parts of the other sheets. Moreover, a proof of any kind should never be regarded as a substitute for carrying out adequate checks at earlier stages, i.e. in the field and on the light table. As has been emphasized in the chapters on survey, every time the scientist makes a check he should be confident that his work will pass it because of simpler and cheaper checks carried out step by step before it. Negative proofs are cheaper than litho-printed proofs, but they are still expensive and the scientist and cartographer should have done everything they can to ensure that as few errors or clashes as possible are revealed by them. Once this has been done and the necessary corrections have been made to the negatives, or if extensive to the fair drawings from which fresh negatives are made, the negatives are ready for the next process – printing.

Before describing printing we should look briefly at some more complex processes which require extra equipment and skill by the photographers and printers. So far we have been describing the production of *line maps* in which all the images, whether of lines, symbols, or lettering, are either completely opaque or completely transparent on positives or negatives so that the printing plate receives ink at full strength for the images and no ink at all in the background. As we shall see in the next section the basic principle of litho-printing is that every part of the printing plate, and consequently of the paper, either receives ink at full strength or not at all; in this sense it is a digital and not an analogue process. Anyone who has looked closely at a picture in a newspaper, preferably with a magnifying glass, will have realized that the different shades in this are produced by a series of very fine dots in a regular pattern, several dots occupying a square millimetre. The original photograph, with its continuous gradation from white to black through varying shades of grey, has been photographed through a *screen* (a very fine mesh of dots in a regular pattern) to produce a *half-tone*. This comes out on the negative as a series of very fine transparent clear dots, and the size of these will depend on the intensity of shade in the area being photographed. Thus the gradation of density is reproduced on the printing plate (and hence on the paper) not directly but by varying the size of the black dots, or they may be of any other colour, without varying the intensity of the ink itself.

There are three common applications of this process in mapping:

(i) to reproduce the grades of hill shading;
(ii) to reproduce scaled copies of the air or satellite photographs as a background to form a photo-map on which names and symbols can be added in full colour (p. 226);
(iii) to reproduce a topographic map as a subdued background to some special features (e.g. soil or vegetation boundaries) which are shown in full strength (see p. 267).

The same process can of course be used for mixing colours, and the beautiful multicoloured maps produced in some atlases or by governments for tourist areas – often with illustrations in the margins – or for geologists carry a very large number of colours and tints. These are achieved by sophisticated and skilful combinations of basic colours and black with the use of screens. The techniques lie on the border line between the photographer and the litho-printer: the first prepares the reproduction material and the latter applies it to the printing plates. No attempt will be made here to describe the processes since they are constantly being improved and simplified so as to reduce the number of basic colours and hence of printing plates required. The processes are obviously ones which the scientist cannot attempt himself, or even understand in detail, unless he is prepared to go on a special cartographic course. They are justified either because, as in a tourist map or atlas, large numbers of copies will be produced for sale, or because the structure being mapped is very complex and the results are of economic importance, as in geological maps. In general the scientist will be producing relatively few copies of his map to illustrate a report, and it is unlikely that his budget will run to such refinements, attractive though they would undoubtedly make his results appear. He should also remember what we said in Chapter 10 (p. 256) – that the attractiveness and complexity of a glossy product are not necessarily related to its information content. The more elaborate a map is, the longer it will have taken to produce and the more people will have had a hand in it with the consequent possibilities that it is out of date and that the subsequent cartographic processes have distorted the accuracy of the original.

Lithographic printing

Lithographic printing consisted originally of creating an image on a porous stone, but nowadays a grained or matt-surfaced metal plate is used. The image is then transferred to successive sheets of paper by pressing them and the plate together. Since printing ink is greasy the process depends fundamentally on the fact that grease and water will not mix. In one process the grained surface of the plate is covered by a mixture of chemicals which hardens on exposure to light. After exposure through the negative of the fair drawing (or the scribed fair drawing itself) the sections of plate under the transparent image will have hardened, but the chemical on the remaining areas will remain soft. The plate is then washed and only the hardened chemical covering the image remains. The rest will be a clean grained surface which will hold a certain amount of water. The plate is dampened and then a greasy substance is rolled onto it which adheres only to the hardened image but not to the dampened remainder. If the plate is then dampened again and greasy ink of the required colour is applied, it will adhere only to the image, and if a sheet of paper is pressed against the plate the image will be transferred to this. Details of the chemicals used vary with different processes and so are not given here; with some the image is etched into the plate and so these can be used for a very much larger number of *impressions* or a longer *run*. The negative (with clear images on an opaque background) or positive (opaque images on a clear background in a different process) may be *right reading* or *wrong reading*, and the printing plate will be the reverse of this. If it is wrong reading then direct transfer to the paper will result in an image the right way round, and this is the simplest (and was the original) form of printing press. However, many modern presses are *offset*: the image is transferred first to a rubber blanket and from this to the paper, thus appearing the same way round as on the printing plate. One advantage of this is that it causes less wear in the image which will consequently last for longer runs; the other is that the image on the plate is the right way round and therefore easier to correct.

Having made his plates the printer will also want to make a *proof* before committing himself to running off the full number of copies on his main presses, and for this he uses a hand-fed *proving press*. This will produce copies which are exactly in the form of the final map, and at this stage not only the usual checks are possible but the cartographers and the scientists can see whether they approve of the colours chosen. Several copies can be made at virtually no extra expense beyond that of producing the first and distributed (by mail if necessary) for comment and check, but here again the process is expensive and any major alterations at this stage will cost a good deal, especially if they involve changes of style and affect other sheets of a series which are well advanced in fair drawing or reproduction. In major projects therefore, as with the negative or dyeline proof, it will be worthwhile producing paper proofs at an early stage of a pilot sheet or mock-up and settling any questions of style before the other sheets progress too far. Colour changes are another matter, and it is quite common for proofs to be run off in a number of different shades before deciding on the colours to be used in the final printing. Most colour proofs carry in the margin a series of squares with one for each colour plate, both as a check that all the plates have been used and also as a guide to the printer in matching the colours for the final printing.

The proving press is normally a simpler, slower, and less automatic version of the main printing press, which is designed to produce a large number of copies in a short time. Clearly, since they will use the same plates, they must both be either direct or offset, and if the final number of copies of a special map is relatively few it may not be worthwhile putting them on the main press but cheaper to run them off on the proving press. Both machines include a series of oscillating rollers for distributing the ink evenly over the whole width of the plate, and a series of large-diameter rollers carrying the plate, the blanket (if it is an offset machine), and either a third roller for the paper or a flat bed for this. In a proving press the paper is usually fed by hand, but in a fast rotary press it will be picked up one sheet at a time by various devices involving grippers and small vacuum tubes. The positions of the plate and of the paper are adjustable, so that once the first colour (normally black) has been printed the others can be registered exactly on it using the corner marks on each to do so and also checking that any sensitive areas (particularly colour-filled roads) are also in register.

When the proofs have been run off, distributed, checked, and returned, and any changes of colour have been agreed and the plates corrected or remade from corrected negatives, printing can proceed. By then the scientist must of course have decided how many copies are required, and in doing this he should be conscious of the economics of the printing process and should have obtained estimates for the different cost of different numbers of colours and of copies. The more complicated and multicoloured is the map, the greater will be the difference between the two parts of this cost which are similar to capital and recurrent expenditure: the 'capital cost' of printing the first copy, and the 'recurrent cost' of each copy thereafter. The cost of fair drawing, reproduction, and printing of the first copy of a map (excluding the cost of survey) can reach £1000 or more; the subsequent cost of each additional copy·may be as little as 10p or less because it has only to cover the cost of the sheet of paper, the small amounts of ink on it, and the time spent in the machine, which may only be a few seconds. Thus any order for copies should be on the generous side and make allowance for future as well as present requirements; it is better to have extra copies lying in store unused for years, if they are likely to be used eventually, than to find a few years later that more copies are required which would certainly involve putting the plates through the machine again, probably re-making them, possibly re-making the negatives, and possibly even re-making the fair drawings or even repeating the whole survey if the originals

FAIR DRAWING, REPRODUCTION, AND PRINTING   305

have not been kept in safe place where they can be found. The more complicated is the map, the more important it is to be generous in ordering, because paper, ink, and machine time in printing extra copies are relatively cheap compared with the skilled work which has gone into making the first copy. It is also worth remembering that if action on a report or map is delayed for several years (as has been known to happen) and it has never been printed, then it may be impossible to locate a copy; but if only a hundred or so copies have been printed then one at least will be available somewhere. Without printing the results and, as we said earlier, permanently marking the survey framework, all the work done by the scientist may be lost as if it had never existed.

GLOSSARY OF TECHNICAL TERMS

Abney level: Small instrument for measuring vertical angles approximately (p. 55).

Acuity: Stereoscopic acuity is the capacity to fuse two stereoscopic photograph images precisely (p. 198).

Aerotriangulation: A system of triangulation connecting minor control points based on the geometry of air photographs (pp. 173, 214).

Alidade: A sight rule with either simple folding sights at its ends or, in the telescopic alidade, a telescope which can be used for more precise sighting and for measuring vertical angles; also that part of a theodolite which is used for sighting (pp. 59, 61, 112).

Anaglyph: A system for viewing a stereoscopic pair of photographs by printing or projecting them in complementary colours and then viewing them through coloured spectacles so that each eye sees only one photograph (p. 213).

Analogue: Data in analogue form are recorded, displayed, or handled by an analogy with the phenomenon being recorded; a common example is the hand of a clock or the pointer of a speedometer (p. 217).

Analytical: An analytical plotter reconstructs a three-dimensional model by computation from simple measurements on two photographs; in general any solution which employs mathematics (p. 220).

Aneroid barometer: Instrument for measuring differences of height by differences of air pressure at different points (p. 56).

Approximate plotter: A stereoscopic plotter using an approximate solution and normally taking paper prints instead of glass or film diapositives (p. 221).

Arrow: A large skewer made of stiff wire with a ring at one end (to which a coloured rag or tape can be tied) and a point at the other, used for marking precisely on the ground the point to which measurements are made with a tape or chain (p. 16).

Astrafoil: A proprietary plastic material in sheet

GLOSSARY OF TECHNICAL TERMS 307

form designed for precise map work where only very small distortions due to changes in temperature or humidity can be accommodated (p. 283).

Astrofix: *See* Fix.

Azimuth: A true bearing or angle from the true north; usually obtained by astronomical observation (pp. 8, 127).

Balplex: A refined form of the Multiplex using an ellipsoidal mirror in the projector to increase the illumination (p. 216).

Base: Side of a triangle in a triangulation whose length has been measured (pp. 25, 118).
In photogrammetry the base line is the line joining the principal points of two successive overlapping vertical photographs of a strip (p. 177).

Basic plot: The original survey, either compiled from field observations or made by photogrammetry from ground or air photographs (p. 263).

Bay: Section of a taped length corresponding to one tape length (p. 120).

Beacon: Structure or lighted device erected over a survey point so that sights can be taken to it from a distance (pp. 26, 129–30, 140).

Bench mark: A marked ground point whose height is measured precisely (pp. 49, 105).

Binary number (bit): A number based on a scale of only two digits (0 and 1) instead of the usual ten (0–9); a bit is one digit of such a number (p. 260).

Blunder: *See* Errors.

Boundary box: A small diagram in the margin of a map showing approximately the positions of international or other boundaries which may or may not be shown on the face of the map (p. 296).

Bridging: *See* Aerotriangulation.

Cadastral: Of surveys or maps relating to property boundaries and rights (pp. 240, 251–3).

Cairn: A survey beacon made of rocks or stones (p. 130).

Calibration: The checking of any instrument or measuring device, such as a camera, to determine its actual, as compared with its theoretical, performance or specification (p. 71).

Case, upper or lower: Upper case, capital letters; lower case, small letters (p. 293).

GLOSSARY OF TECHNICAL TERMS

Catenary (taping): Measuring a distance precisely with a tape hanging free between two or more supports in the shape of a chain (Latin *catena*) of very small links (p. 120).

Centring: Placing an instrument precisely over a ground mark or over a previous and subsequent target position; forced centring: a system of interchangeable tripods on which the targets and the instrument are forced to occupy the same points successively (p. 111).

Chain surveying: Surveying by distance measurement only, originally using a chain but also applicable to a tape or other means of direct distance measurement; step chaining: measuring distances along a slope with one end of the chain or tape touching the ground and the other elevated so that the length of it is horizontal (pp. 16, 33).

Change point: Point at which a levelling staff is held fixed while the instrument moves between two successive stations (p. 49).

Circle (great or small): A circle on the earth's assumed spherical surface; a great circle has its centre at the centre of the earth, whereas a small circle does not (p. 243).

Clinometer (usually Indian): Instrument for measuring vertical angles, used particularly with a plane table (p. 59).

Close, closure, misclosure: A triangle or other geometrical figure is closed when all its angles have been measured and the difference, or misclosure, between their actual and theoretical sums is known; a triangulation chain or a traverse is closed between two fixed points at its ends (pp. 26, 36, 46, 152-3).

Collimation: Collimation error, lack of perpendicularity between the axis of a telescope, as in a theodolite, and its axis of rotation (p. 109); collimation marks: marks at the sides or in the corners of a camera format, and hence appearing on photographs taken with it, for defining the position of the principal point (pp. 76, 170).

Colour separation: The construction of separate fair drawings for each of the different basic colours required in a printed map for which a

GLOSSARY OF TECHNICAL TERMS 309

	separate printing plate will be required (p. 266).
Compass, prismatic:	Magnetic compass in which the bearing of an object is read through a prism simultaneously with sighting on the object (p. 23).
Compilation diagram:	A small diagram in the margin of a map showing the material from which it was compiled (p. 296).
Continuous revision:	The regular up-dating of basic plans, in contrast to periodic revision in which they are only updated at intervals of several years (p. 247).
Contour flying:	Special flying with aircraft carrying geophysical instruments so that these are kept at a constant height above the ground (p. 234).
'Cook':	*See* Fudge.
Copy:	An original (p. 299).
Corner marks:	Small right-angled marks in each corner of a fair drawing required for exact register of this with others in cases where the map will be printed in more than one colour (pp. 266, 290).
Correction:	The small amount required to bring a measured quantity to its correct value, of opposite sign to the error due to the same cause (p. 123).
Correction, altitude or sea level:	The correction required to convert an actual horizontal distance to its equivalent at sea level (p. 124).
Correction, sag:	*See* Sag.
Correction, slope:	The difference between a distance measured along a slope and the equivalent horizontal distance (p. 124).
Correction, temperature:	*See* Standard.
Countersection:	Fixing a point by one bearing taken to it from a fixed point and a cross bearing taken to a second fixed point from the point being fixed (p. 65).
Crab:	The horizontal angle between the direction in which an aircraft is pointing and that in which it is actually flying due to the effect of a cross wind (p. 170).
Cross hatching:	Two patterns of rulings drawn or printed at a substantial angle to each other (usually $90°$ or $45°$) (p. 294).
Derived maps:	*See* Second-generation maps.

GLOSSARY OF TECHNICAL TERMS

Detail: Natural or man-made features shown on a map in plan only (p. 9).

Detail, hard: Features with definite sharp or hard outlines (usually man-made) from which precise measurements can be made (p. 10).

Detail point: Points on a feature which have been fixed by measurement, with the detailed shape of the feature being interpolated between them (p. 19).

Detail, soft: Indefinite features, usually natural ones (p. 10).

Discritical mark: A small sign (dot, comma, etc.) added above or below a letter to show that it has a special pronunciation (p. 278).

Diapositive: Positive copy of a photograph, which may be enlarged or reduced, on transparent material such as glass or film (p. 203).

Digital: Data in digital form is recorded in discrete units which can be written or displayed as figures or recorded electronically (p. 260).

Digital encoder: A device for converting movements of a measuring instrument into numbers (p. 219).

Distance angle: The small angle of a thin triangle whose measurement determines the distance of that apex from the opposite side (p. 116).

Distortion, camera: Differences in the direction of rays coming from the lens to the format from the directions of the same rays as they enter the front of the lens (p. 169).

Distortion, cartographic: Changes in the dimensions of cartographic material due to heat or humidity (p. 283).

Distortion, height: Differences in the positions on a vertical air photograph of the tops of tall objects and of the ground beneath them (p. 170; Plate 5).

Diurnal wave: Regular daily variation in the air pressure at one point (p. 58).

Dropline contours: Contours defined by the changing width of a cross section (p. 225).

Dumpy level: Spirit level without an elevating screw (p. 43).

Dyeline: A photographic copying process for copying positives onto specially coated paper (pp. 263, 298).

Eccentricity: The small difference in position between the two centres, of rotation and graduation, of the circle of a theodolite (p. 110).

Electromagnetic distance measurement (EDM): A system of measurement of distances by

GLOSSARY OF TECHNICAL TERMS 311

	timing the passage of an electromagnetic signal between two objects (either two transmitter/receivers or one transmitter/receiver and a reflector); frequencies vary from those of radar through microwave to infra-red and optical (pp. 121–2).
Errors:	Differences from a correct value (p. 34).
Errors, absolute:	Errors of position relative to a grid or graticule (p. 80).
Errors, accidental:	Small, unbiased, and unavoidable errors in observations (pp. 34, 77).
Errors, gross:	Large errors, mistakes, or blunders e.g. mis-readings (pp. 35, 79).
Errors, relative:	Errors of position relative to adjacent features or framework point (p. 80).
Errors, systematic:	Small, biased, and sometimes avoidable errors in observations (pp. 34, 78).
Fair drawing:	The conversion of a basic plot to professionally drawn and lettered material from which a printed map can be produced by photo-mechanical means (p. 263).
Field completion:	Completion in the field of a map made from air photographs to show names and other items not visible in the photographs (p. 274).
Fix:	Verb: to locate the position of a point precisely on the ground; noun: such a point, as in astrofix (a point fixed by astronomical observations (p. 10).
Floating spot (or dot):	A spot of light or a small dark image formed in a stereoscopic measuring device in conjunction with a three-dimensional model by means of which measurements in three dimensions can be made when it appears to touch the model (pp. 191, 199).
Focal length:	The length of the perpendicular from the lens (strictly its rear node) of a camera to the format (pp. 71, 168).
Focal plane shutter:	A camera shutter working at the format near the film and not between the two parts of the lens; it consists of a slit in a blind drawn rapidly across the format thus exposing different parts of it at different times, and not simultaneously (pp. 169, 200)
Forced centring:	A device which ensures that a theodolite and a target placed in succession on a tripod occupy exactly the same spot (p. 111).

312   GLOSSARY OF TECHNICAL TERMS

Format:	The rectangular plane surface in a camera in which the film theoretically lies when being exposed (pp. 71, 170).
Form lines:	Approximate contours (pp. 184, 273).
Framework (Framework points):	Survey structure whose shape, size, orientation, and position have been defined by relatively precise measurement, from which detail and contours can be mapped by less precise and quicker techniques (p. 19).
French curve:	A piece of stiff material (wood or plastic) with varying curves used for steadying a pen when drawing curved features (p. 286).
'Fudge':	To adjust observations incorrectly, or even fraudulently, so as to reduce the misclosure of a set of observations to an acceptable figure (pp. 52, 118).
Fusion:	The skill or act of causing the two different images of an object recorded on an overlapping pair of photographs to fuse into one three-dimensional image (p. 194).
Gazetteer:	A list of geographical names with their co-ordinates, or with some other means of locating them on maps (p. 244).
Grain:	The addition of a matt surface to a smooth material so that ink will adhere to it (p. 283).
Graticule:	Lattice formed by parallels and meridians of latitude and longitude (p. 243).
Grazing ray:	A line of sight passing near a building, rock, or other object whose temperature difference from that of the air will cause the ray to be bent (p. 139).
Grid:	System of straight lines at right angles laid out precisely on the ground or on a map, usually in squares (pp. 20, 155-6, 244-5).
Grid convergence:	The angle between grid north and true north (p. 245).
Grid zone:	A section of a grid system of wide extent, usually bounded by two meridians several degrees apart (p. 244).
Ground control:	A series of points fixed by ground survey and identified on air photographs which are used for controlling a map made from these photographs (pp. 184-7, 226-34).
Ground mark:	A mark inserted in the ground to mark precisely and unambiguously, and if possible permanently, the site of a survey station;

GLOSSARY OF TECHNICAL TERMS 313

	a precast ground mark is cast in concrete and then put in the ground; a cast-in-place ground mark is cast in concrete into a prepared hole in the ground (pp. 105–7).
Hachures:	Lines of varying thickness drawn down a slope on a map to indicate its steepness and direction (p. 273).
Hair line:	Thin line in a letter which is liable not to reproduce in printing (p. 293).
Halation:	The expansion of light images on a photographic positive, or of dark images on a negative (p. 230).
Half-tone printing:	A technique of printing using screens (see Screens).
Helio (Heliograph):	A system of mirrors for reflecting the sun and directing a narrow intense beam of light towards a distant observer (p. 130).
Hill shading:	A technique for showing relief by shading the slopes away from a supposed source of light above the northwest corner of a map (p. 273).
Identification (photo-):	The precise determination of the *position* of the image of a ground point or feature on an air photograph or map (pp. 228–33).
Impression:	The process of printing one colour onto one sheet of paper; commonly used by printers as a unit of printing work (p. 303).
Interpretation (photo-):	The determination of the *nature* of a feature seen on air photographs (pp. 202–4).
Intersection:	The fixing of a point by bearings or rays to it from at least two points which are already fixed (p. 29).
Italics:	Sloping letters (p. 293).
Kelsh:	A form of projection stereoscopic plotter in which full-sized diapositives are used with the relevant sections illuminated by spot lights (p. 216).
Leg (of a traverse):	*See* Traverse.
Legend:	A descriptive list in the margin of a map giving details of the symbols etc. shown on its face (pp. 269, 296).
Letterpress:	Names or words printed on a material suitable for fixing to a fair drawing (p. 290).
Levelling:	The measurement of heights (p. 41).
Levelling, spirit:	Measuring heights with a telescope made horizontal by a spirit level and observing to a graduated staff (pp. 41, 52).
Levelling, tacheometric:	*See* Tacheometry.

GLOSSARY OF TECHNICAL TERMS

Levelling, trigonometrical (or heighting).: Measuring relative heights by observing the vertical angle between two objects and calculating the difference in height from this and the known distance by trigonometry (p. 55).

Light: An illuminated survey beacon, either by the sun using mirrors, or by a signalling lamp (p. 129).

Light table: A special table with a glass window so that drawings or other materials can be illuminated from below (p. 285).

Line map: A map in which the detail has been drawn by hand and not reproduced directly from air photographs (p. 221).

Lingua franca: A common language (e.g. Swahili) used over a wide area in addition to local languages (p. 279).

Lithography (printing): The technique of printing by means of a greasy ink image on a stone or (nowadays) a metal plate (pp. 302-3).

Magnetic north (or south): The direction in which a magnetic needle points (pp. 24, 295).

Maps, first generation: Maps compiled directly from original surveys (p. 239).

Maps, second generation: Maps compiled (usually on a smaller scale) from other maps (pp. 239, 254-7).

Mean: The average of a number of measurements of the same quantity (p. 78).

Mereing: Defining a public administrative boundary (e.g. between two parishes or electoral districts) in the United Kingdom (p. 247).

Meridian: A great circle on the earth's surface passing through both poles, or its representation on a map (p. 243).

Micrometer: A device for interpolating accurately between two graduations of a scale by the rotation of a graduated drum actuating a moving mark; one rotation of the drum makes this cover the distance between two graduations on the scale (p. 113).

Minor control points: Small points of detail in the corners of an air photograph or an overlap used for assembling a series of photographs into their correct relative positions to form a framework within which detail can be plotted (p. 173).

Minute of arc: A small angle in which the length perpendicular to the arms of the angle is about 1 part in 3000 of their length (p. 56).

GLOSSARY OF TECHNICAL TERMS 315

Misclosure: *See* Closure.
Mock-up: An experimental rough model used in planning the layout of a map (p. 295).
Model: The three-dimensional image formed by the correct projection or viewing of an overlapping pair of photographs (p. 190).
Mosaic: An assembly of air photographs which have been cut and stuck together to form a continuous picture; in a˙controlled mosaic the photographs have been scaled and positioned, and in an uncontrolled mosaic they have not (p. 225).
Multiplex: A type of projection stereoscopic plotter in which a pair of or several projectors can be set up over a table, each projector taking a reduced diapositive of the original photographs (p. 212).
Neat line: The edge of a photograph or overlap, or of the detail on a map, i.e. excluding the marginal information (pp. 189, 295).
Negative: Material with transparent or white images on an opaque or dark background (pp. 265, 283).
Negative proof: Proof of a multicoloured map made from negatives of the fair drawings and not from printing plates (p. 299).
Node: Point at the front or back of a lens into which rays of light converge before passing through the lens (p. 71).
Oblique (air photograph): Air photograph taken with the camera axis making a substantial angle with the vertical; in a high oblique the horizon appears, and in a low oblique it does not (p. 163).
Offset: A printing process where the original image is not transferred directly from the plate to the paper but via a rubber blanket so that it is the same way round as on the plate and the image on the plate lasts longer (p. 303).
Opaque beacons Beacons or targets seen by light reflected from them (p. 132).
Optical square: A device for setting out right angles or straight lines using prisms for viewing two objects simultaneously (p. 249).
Orientation: Making a map or field sheet, as on a plane table, parallel to the ground (p. 61).
Orientation, absolute: In photogrammetry, making the model formed by the overlap of two photographs fit to four points in its corners whose

Orientation, relative: ground positions and heights are known (p. 227).
Making the two overlapping photographs of a stereoscopic pair form a perfect model in the area of the overlap (p. 227).

Orthophotograph: A copy of an air photograph from which height and tilt distortions have been removed in a three-dimensional plotting machine so that it is all at uniform scale (pp. 193, 222).

Overlap: The area common to two vertical air photographs (pp. 162, 189).

Overlap, forward: The area common to two successive air photographs of a strip (pp. 161-2).

Overlap, lateral: The area common to two air photographs of adjacent strips (pp. 161-2).

Overlay: A sheet of transparent material with special information printed or drawn on it for use in conjunction with a basic topographic map (p. 267).

Overprinting: The printing of similar information in a special colour onto such a map, which may itself be printed in a subdued colour (p. 267).

Ozalid: See Dyeline.

Parallactic angle: The angle subtended at a point by the lines joining it to the two positions from which it was observed (p. 196).

Parallax: The apparent movement of one object relative to another at a different distance from the observer due to a change in his position; x parallax: the small distances parallel to the base line of two overlapping photographs by which the images of distant objects are different on the two photographs; y parallax: the same distances at right angles to the base line of the two photographs (p. 206).

Parallax bar: Device for measuring relative heights by the measurement of parallax differences on an overlapping pair of photographs (p. 206).

Parallel (of latitude): A circle on the earth's surface with its centre on the axis of rotation joining the two poles, and its plane perpendicular to this; or its representation on a map (p. 243).

Pass points: See Minor control points.

PCGN: The British Permanent Committee for Geographical Names for British official use (p. 280).

GLOSSARY OF TECHNICAL TERMS 317

Pecked line: A line broken into short lengths or dashes (p. 270).
Permatrace: *See* Astrafoil; this material is less liable to shatter if dropped.
Phase: The effect of asymmetrical lighting, as in the phases of the moon (p. 132).
Phase error: The error caused by sighting on an asymmetrically illuminated target such as a pillar with the sun shining on one side (p. 132), or a graduation on a metal scale (p. 114).
Photogrammetry: The technique of making measurements (including maps) from photographs (Chapters 7-9).
Photo-index: A small-scale map purporting to show the layout of aerial photographs (p. 165).
Photo-map: A map in which the detail has been reproduced directly from air or satellite photographs, although some features may be emphasized by lines, lettering, or other symbols (p. 226).
Photo-theodolite: Combination of a calibrated camera with a theodolite for taking precise terrestrial photographs from ground stations (p. 76).
Pixel: A unit of resolution in satellite imagery transmitted by radio (p. 258).
Plane table: A flat drawing board fitted to a tripod so that it can be used in a horizontal position and either rotated about a vertical pivot or clamped in one position (p. 58).
Planimetric map: A map without contours (p. 22).
Plot: A field sheet or photogrammetic compilation in its original form before being copied or fair drawn (p. 174).
Plumb bob: A heavy weight with a point underneath on the end of a string, used for centering an instrument accurately over a ground point (pp. 59, 110).
Plumb point: The point vertically below the camera when an air photograph is taken (p. 170).
Point: Unit of height for letters to indicate their size (p. 293).
Positive: Material with opaque images on a white or transparent background (pp. 265, 283).
Premark: A ground point marked (usually in white or yellow) so that it will appear on subsequent air photographs (p. 233).
Principal point: The point where the lens axis of a camera meets the format; the position of this point

318 GLOSSARY OF TECHNICAL TERMS

	on a photograph, and sometimes its corresponding position on the ground in the case of a vertical air photograph (pp. 71, 170).
Printing down frame:	A large printing frame used for copying material photographically at the same scale by contact (pp. 265, 298).
Process camera:	A large precise camera used for copying fair drawings or other material at the same or a different scale (pp. 265, 298).
Projection, map:	A system for representing the assumed spherical earth's surface on a flat piece of material such as a map (p. 155).
Projection, conformal:	Projection in which the local shapes and angles are preserved but the scale varies (p. 155).
Projection, equal-area:	Projection in which areas are preserved but at the expense of shapes (pp. 155, 242).
Projection, orthogonal:	Projection where the centre of projection is infinitely far away (p. 167).
Projection, perspective:	Projection where the centre of projection (as in a camera) is relatively near to the plane onto which the images are being projected (p. 167).
Projection plotter:	A stereoscopic plotting instrument in which the rays of light reaching the camera are reconstituted by light sources projecting images of the photographs onto a table (p. 212).
Projection, Transverse Mercator:	Projection used frequently for large-scale and medium-scale maps and for computing frameworks (p. 155).
Projection, Universal Transverse Mercator:	World-wide Transverse Mercator projection between $80°N$ and $80°S$ covering the whole world in $6°$ belts of longitude (pp. 155-6).
Proof:	An advance printed copy normally on paper used for minor corrections and checking colours (p. 303).
Pseudoscopic:	A false stereoscopic appearance of a three-dimensional model formed either by viewing different objects, or by viewing an object that has moved at different times, or (in photogrammetry) by reversing a stereoscopic pair of air photographs to that heights appear as depths and vice versa (p.197).

GLOSSARY OF TECHNICAL TERMS 319

Quadrilateral:	A figure with four straight sides (p. 26).
Quadrilateral, braced:	A quadrilateral in which all the angles, including those between the diagonals and the sides, have been measured (p. 26).
Radar altimeter:	An altimeter which measures by radar reflections from the ground the height of an aircraft above it (p. 235).
Radial (line plotter):	In a direction from the principal point of a vertical air photograph; a plotter using this principle (pp. 170, 182-4).
Range:	Difference between the largest and smallest observations of the same measurement (p. 78).
Range finder:	A device, used mainly by the military, for measuring distances to an object by bringing into coincidence two images of it sighted from the ends of a bar of fixed length (p. 151).
Rectification:	The production of a uniformly scaled copy of an air photograph of flat ground by removal of the tilt distortions (p. 222; Plate 3).
Redundant observation:	An observation which acts as a check on others used for determining the positions or shapes of a survey framework (p. 10).
Reference object (RO):	The object or target to which the first (and usually also the last) sighting is made in a round of angles (pp. 116, 133).
Refraction:	The bending of light rays in a transparent medium such as air or water due to their varying densities e.g. near the ground which may be hotter or cooler than the air (p. 139).
Register:	The exact fitting of two or more images in different colours on a multicoloured map (p. 264).
Relief:	The shape of the ground (pp. 272-3).
Reproduction, photographic:	The reproduction of an original or fair drawing by photographic means (p. 263).
Reproduction, photo-mechanical:	The reproduction of an original or fair drawing by a combination of photographic and mechanical means (p. 263).
Resection:	The fixing of a point by magnetic bearings from it to at least two fixed points, or by observing the angles at it between rays to at least three points that are already fixed (pp. 29, 66).
Road pen:	A pair of ruling pens coupled together so that

320 GLOSSARY OF TECHNICAL TERMS

Ruling:	two lines of constant width can be drawn a constant distance apart. A pattern of parallel lines very close together (p. 294).
Ruling pen:	A special pen designed for drawing lines of constant width (p. 286).
Run (long or short):	The number of printing impressions or copies of a publication (long = large; short = small) (p. 303). Also the distance traversed by micrometer hairs with one complete cycle of the graduated scale (p. 113).
Sag:	The difference from a straight line of a tape suspended in catenary between two supports; sag correction: the difference between the suspended and the straight line lengths (pp. 123-4).
Satellite imagery (or photographs):	Information in pictorial form obtained, on various wavelengths, of the earth's surface by an orbiting satellite and transmitted to the earth by various techniques (pp. 257-62).
Satellite station:	A survey station near a beaconed survey point which is occupied to save dismantling and re-erecting the beacon (p. 132).
Scale (or representative fraction):	The ratio of distances measured on a map or air photograph to their equivalents on the ground (pp. 16, 166, 295).
Scale factor:	The ratio of the scale of a grid at any point to its value in the area or areas where it is unity. Scale error is the difference between the scale factor and unity (pp. 124-5).
Screen:	A sheet of transparent material carrying a very fine and regular pattern of dots by which a continuous-tone picture can be broken down into a half-tone picture in which variations of intensity are represented by variations in the size of the dots and not in their colours or shade (p. 301).
Scribing:	The technique of fair drawing in which the lines and symbols are cut out of an opaque medium instead of being drawn with a pen in an opaque ink on a transparent or white medium (pp. 265, 289-90).
Second of arc:	A very small angle in which the length of the perpendicular to the arms of the angle is about 1 part in 200 000 of their length (p. 108).

GLOSSARY OF TECHNICAL TERMS 321

Serif:	*See* Hair line.
Setting out:	Putting on the ground what is on paper (p. 20).
Sextant:	An instrument for observing the angle between two objects from a moving vehicle by the simultaneous observations of one object directly and the reflection of the other in two mirrors, one of which rotates against a graduated scale of degrees (p. 91; Fig. 4.2).
SIM:	*See* SUSI, SIM.
Slotted template:	A stiff square of plastic representing a vertical air photograph, with a hole at the centre representing the principal point and slots radiating from this to the positions on the photograph of the preceding and following principal points and of the minor control points (pp. 180-2).
Space rod:	An accurately machined rod representing the ray of light from an object to the lens of a camera; by using two in a stereoscopic plotting machine the position of a point whose image appears on both photographs can be reconstructed mechanically (p. 217; Plate 2).
Spherical excess:	The amount by which the sum of the three angles of a triangle drawn on a sphere exceeds $180°$ (p. 153).
Spotting camera:	Small camera used for taking vertical photographs (usually on 35 mm film) to locate positions along the trace of an aircraft carrying geophysical instruments (pp. 234-5).
Stadia hairs:	Short hairs in the diaphragm of a telescope which subtend a fixed angle (usually 1/100 at the eyepiece (pp. 45, 143, 150-1).
Standard:	The length of a measuring tape or base used for calibrating a field tape (p. 123).
Standard temperature (tension):	The value of temperature (tension) at which a tape's length has been calibrated (p. 123).
Station pointer:	Hydrographic plotting device with three straight arms the outer two of which can rotate about a central pivot and against a circle graduated outwards in degrees from the zero of the central arm so that a point's position on a plotting board can be plotted from the two angles between the three rays observed from it to three fixed points (p. 93).

322 GLOSSARY OF TECHNICAL TERMS

Stereoscope: A device for viewing two photographs so as to produce a three-dimensional image (p. 198; Plate 1).

Stereoscopic (principle, plotter): The use of two overlapping views of an object seen with both eyes to form a three-dimensional model whose size and shape can be measured either directly in the brain or indirectly by looking at two photographs of the object taken from different points; a plotter using this pinciple (pp. 193-202, 211-21; Plates 2 and 4).

Stipple: A pattern of fine dots used to represent buildings, sand, etc. (pp. 247, 275).

Strip (of air photos): A succession of air photographs taken on a single line of flight (p. 161).

Subtense (bar or base): The use of a small angle subtended at the observer to measure the distance from a bar of fixed length or a short measured base between two targets (pp. 146-7).

Supralap: The strip across the middle of a vertical air photograph covered by both the preceding and following photographs of the same strip (pp. 161, 176).

SUSI, SIM: Supply of Unpublished Survey Information; Supply of up to date Survey Information on Microfilm. Both are systems used by the Ordnance Survey for supplying up to date copies of master traces of their large-scale plans, the former at their own offices and the latter through commercial agents (pp. 247-8).

Tacheometry (or Tacheometric levelling): Technique for measuring distance and height differences normally by optical means using a theodolite with stadia hairs and observing the distance intercepted by them on a graduated staff held vertically (pp. 55, 143, 150-1).

Temperature correction: *See* Standard temperature.

Thalweg: The line following the deepest part of a stream (p. 281).

Thematic maps: Maps showing themes or features other than topographic ones or legal boundaries e.g. population distribution (p. 239).

Theodolite: An instrument for observing precise angles in the horizontal and vertical planes (pp. 107-9).

GLOSSARY OF TECHNICAL TERMS 323

Tick (graticule or grid):	A short line in the margin of a map to mark the exact position of one end of a graticule or grid line where this is not shown in full on the surface of the map (p. 245).
Tidegauge:	A device for measuring the height of the tide (p. 94).
Tilt:	The angle between the axis of an aerial camera and the vertical; more particularly when measured at right angles to the direction of flight (pp. 166, 213).
Tip:	The tilt of a camera axis in the direction of flight (p. 213).
Topographic maps:	Maps showing the shape of the ground and the natural and man-made features on it (p. 239).
Traverse:	A series of straight lines between successive points on the ground at which their bearings (or the angles between them) have been measured, and between which the distances have been measured; traverse leg, one such line (pp. 30-1, 134-8, 157-8).
Triangle of error:	The triangle produced when three rays or bearings of an intersection (or more specifically of a resection) do not meet at a point (p. 68).
Triangle, thin:	Triangle in which one angle is small (p. 25).
Triangle, well-conditioned:	Triangle in which no angle is less than $30°$ (p. 26).
Triangulation:	The establishment of a survey framework by the measurement of horizontal angles with the size or scale established by only one or two measured distances (pp. 22-30, 126-34, 152-6).
Tri-camera or Trimetrogon (photography):	A system of aerial photography in which three cameras are used, one vertical and the other two pointing obliquely (usually at $60°$) from the vertical and at right angles to the direction of flight; Trimetrogon refers to the system used by the US Air Force in World War II in conjunction with the US Geological Survey, with cameras fitted with Metrogon lenses (p. 163).
Trig:	Short for both triangulation and trigonometrical.
Trig diagram:	A diagram showing the proposed or actual rays observed in a triangulation (p. 126).

Trig heights: Heights obtained by trigonometry from a known distance and vertical angle (pp. 55, 140).

True north: The direction of the North Pole (along a meridian) (p. 295).

Vacuum back: A device for flattening film or other flexible material into a plane for accurate and sharp photographic exposure or copying (pp. 169, 298).

Vernier: A device for interpolating between the divisions on a graduated scale by matching it with a section of similar scale in which the divisions are different by a fixed proportion so that the coincidence of divisions between the two scales indicates the interpolated value (p. 113).

Vertical interval (VI): The vertical interval between successive contours (p. 272).

Vinten (camera): *See* Spotting camera.

Zero: Pre-set reading on the horizontal circle of a theodolite which will only rarely be near zero on its scale; other values are spaced at regular intervals round the circle to eliminate any errors in its graduation (p. 116).

RECOMMENDED READING

Allan, A. L., Hollwey, J. R., and Maynes, J. H. B. (1968). *Practical field surveying and computations*. Heinemann, London. Advanced but needs updating on computations. [Chapters 5 and 6].
Anon. (1965). *Admiralty manual of hydrographic surveying*, Vol. I. HMSO, London. [Chapter 4].
— (1966). *Manual of photogrammetry*, 3rd edn. American Society of Photogrammetry, see section on tri-camera photography. [Chapter 7].
— (1975, 1979). *Proc. Conf. of Commonwealth Surveyors*. Overseas Development Administration, London. Not too technical; management-oriented survey papers.
— (1935). *Hints to travellers*, 11th edn. Vol. I, pp. 159–64. Royal Geographical Society, London. [Chapter 7.]
Atkinson, K. B. (ed.) (1980). *Developments in close-range photogrammetry*. Applied Science Publishers, London. For non-topographical applications. [Chapter 9.]
Berrangé, J. P. (1975). Advice for small expeditions to forested regions. *Geogr. J.* **141**, 421–9. Also available as a pamphlet from the Royal Geographical Society, London. [Chapters 7, 9 and 10.]
Blachut, T. J., Chrzanowski, A., and Saastamoinen, H. J. (1979). *Urban surveying and mapping*. Springer Verlag, New York. Advanced textbook. [Chapters 5, 6, 9, and 10.]
Bomford, A. G. and Paterson, W. S. B. (1958). The survey of South Georgia. *Emp. Surv. Rev.* **14**, 204, 242. Very good on exploratory survey. [Chapters 3 and 6.]
Burnside, C. D. (1971). *Electro-magnetic distance measurement*. Crosby Lockwood, London. New edition in hand. [Chapter 5.]
— (1979). *Mapping from aerial photographs*. Crosby Lockwood, London. Advanced but comprehensive. [Chapters 7–9.]
Coleman, A. and Shaw, J. E. (1980). *Land Utilization Survey: Field Mapping Manual*. Second Land Utilization Survey of Britain, King's College, London [Chapter 11.]
Cooper, M. A. R. (1971). *Modern theodolites and levels*. Crosby Lockwood, London. New edition in hand. [Chapters 3 and 5.]
— (1974). *Fundamentals of survey measurement and analysis*. Crosby Lockwood Staples, London. Rather advanced. [Chapter 6.]
Crone, D. R. (1963). *Elementary photogrammetry*. Edward Arnold, London. [Chapters 7–9.]
Cross, P. A. and Webb, J. P. (1980). Instrumentation and methods for inertial surveying. *Chart. Land. Surv./Miner. Surv.* **2** (2), 4–27. An account of the latest techniques (advanced). [Chapter 6.]
Dale, P. F. (1976). *Cadastral surveys within the Commonwealth*. HMSO, London. [Chapter 10.]

RECOMMENDED READING

Edwards, D. A. and Partridge, C. A. (eds.) (1977). *Aerial archaeology*, Vol. I. Committee for Archaeological Air Photography, Hertford. See section on archives of aerial photographs. [Chapter 7.]

Garner, J. B., James D., and Bird, R. G. (1976). Surveying. *Estates Gazette*, London. [Chapters 2-6.]

Geelan, P. J. M. (1973). The collection of place names by small expeditions. *Geogr. J.* **139**, 104-6. Also available as a pamphlet from the Royal Geographical Society. [Chapter 11.]

Harley, J. B. (1975). *Ordnance Survey maps*. Ordnance Survey, Southampton. [Chapter 10.]

Ingham, A. E. (1974). *Hydrography for the surveyor and engineer*. Crosby Lockwood Staples, London. [Chapter 4.]

Irvine, W. H. (1974). *Surveying for construction*. McGraw Hill, New York. Good for details of instruments. [Chapter 2.]

Keates, J. S. (1973). *Cartographic design and construction*. Longmans, Harlow, Essex. Highly recommended. [Chapters 11 and 12.]

Kilford, W. K. (1973). *Elementary air survey*. Pitman, London. [Chapters 7-9.]

Loxton, J. (1980). *Practical map production*. John Wiley, New York. Recommended. [Chapter 12.]

Mackie, J. B. (1978). *The elements of astronomy for land surveyors*, 8th edn. Charles Griffin, London. [Chapter 1.]

Mason, K. M. (1927). The stereographic survey of the Shaksgam. *Geogr. J.* **70**, 342. [Chapter 3.]

Miller, K. (1977). *Simple surveying techniques for small expeditions*. Royal Geographical Society, London. [Chapters 2-4.]

Monkhouse, F. J. and Wilkinson, H. R. (1963). *Maps and diagrams*. Methuen, London. Mainly on thematic maps. [Chapters 11 and 12.]

Nørlund, N. E. and Spender, M. A. (1935). Some methods and procedures developed during recent expeditionary surveys in South East Greenland. *Geogr. J.* **86**, 317. Still relevant today for expeditions. [Chapters 3 and 6.]

Olliver, J. G. and Clendinning, J. (1978). *Principles of surveying*. Vol. I, *Plane surveying*. 4th edn. Van Nostrand and Reinhold, New York. [Chapter 2.]

— and — (1980). *Principles of surveying*. Vol. II, *Photogrammetry, adjustment and field astronomy*. Van Nostrand and Reinhold, New York. Advanced. [Chapters 1, 7-9.]

Pugh, J. C. (1979). *Surveying for field scientists*. Methuen, London. [Chapters 2-6.]

Ritchie, W., Tait, D. A., Wood, M., and Wright, R. (1977). *Mapping for field scientists*. David and Charles, Newton Abbot, Devon. [Whole book.]

Robbins, A. R. (1976). Field and geodetic astronomy. In *Military engineering*, Vol. 13, Part 9. Ministry of Defence, London. [Chapter 1.]

Robinson, A. H. and Petchnick, B. B. (1976). *The nature of maps*. University of Chicago Press, Chicago, ILL. [Chapters 10-12.]

Rushworth, W. D. (1975). Electronic calculators for expedition surveyors. *Geogr. J.* **141**, 72-5. Also available as a pamphlet from the Royal Geographical Society, London. [Chapter 6.]

Smith, J. R. (1970). *Optical distance measurement*. Crosby Lockwood London. [Chapter 6.]

— (1973). *Desk calculators*. Crosby Lockwood, London. [Chapter 6.]

Thompson Morris, M. (1979). *Maps for America*. US Geological Survey Washington, DC. A description of US Geological Survey maps. [Chapter 11.]

Verstappen. H. Th. (1977). *Remote sensing in geomorphology*. Scientific Publications, Comprehensive but expensive. [Chapters 7, 8, and 10.]

White, L. P. (1977). *Aerial Photography and remote sensing for soil survey*, Oxford University Press. [Chapter 7.]

Wolf, P. R. (1974). *Elements of photogrammetry*. McGraw Hill, New York. [Chapters 7-9.]

Wright, J. W. (1939). Survey on polar expeditions. *Polar Rec.* **3**, 144-68. Detail survey and terrestrial photography. [Chapters 3 and 6.]

— (1945). War time exploration with the Sudan Defence Force in the Libyan Desert 1941-3. *Geogr. J.* **105**, 100. For desert traversing. [Chapter 10.]

— (1948). A simple form of catenary apparatus for minor control traverses in rough country. *Emp. Surv. Rev.* **9**, 344. [Chapter 5.]

— (1950). The spelling of African place names. *Emp. Surv. Rev.* **10**, 284. Statement of principles. [Chapters 10 and 11.]

— (1951). Reconnaissance mapping from Trimetrogon air photographs in the Anglo-Egyptian Sudan. *Emp. Surv. Rev.* **9**, 2. Comprehensive account of simple techniques. [Chapter 7.]

— (1955). The triangulation of cultivation overlooked by high ground. *Emp. Surv. Rev.* **13**, 1-7, 51-58. Solving a special problem. [Chapter 6.]

— (1973). Air photographs for small expeditions. *Geogr. J.* **139**, 312-322. Good for finding where photographs are available. Also available as a pamphlet from the Royal Geographical Society, London. [Chapter 7.]